J. WESTON
WALCH
PUBLISHER
Portland, Maine

EASY
Science Demos & Labs

Life Science

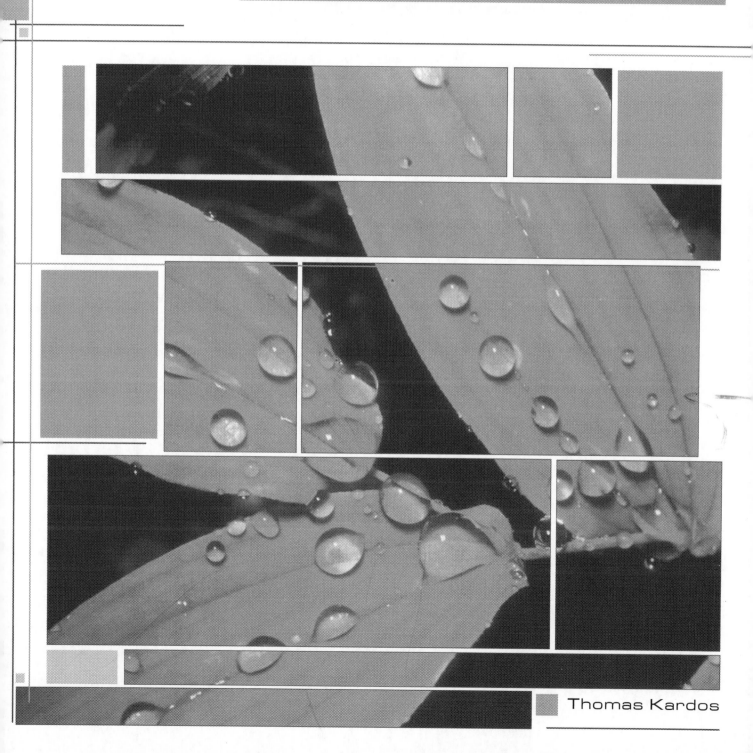

Thomas Kardos

User's Guide
to
Walch Reproducible Books

Dedication

This book is dedicated to my darling wife, Pearl, who throughout this project assisted me with great patience. As a nonscience educator, she helped me develop this book into an easy-to-use and comprehensible resource.

1   2   3   4   5   6   7   8   9   10

ISBN 0-8251-4501-5

Copyright © 1996, 2003
J. Weston Walch, Publisher
P.O. Box 658 • Portland, Maine 04104-0658
walch.com

Printed in the United States of America

Contents

Demos and Labs

Appendix

Preface

As a middle school teacher, many times I found myself wishing for a quick and easy demonstration to illustrate a word, a concept, or a principle in science. Also, I often wanted a brief explanation to conveniently review basics and additional information without going to several texts.

This book is a collection of many classroom demonstrations. Explanation is provided so that you can quickly review key concepts. Basic science ideas are hard to present on a concrete level; the demonstrations fill that specific need. You will also find 10 specially created laboratory activities for middle school students that are safe enough for young people to do on their own. These labs add a deeper level of understanding to the demonstrations.

An actual teacher demonstration is something full of joy and expectation, like a thriller with a twist ending. Keep it that way and enjoy it! Try everything beforehand.

We need to support each other and leave footprints in the sands of time. Teaching is a living art. Happy journey! Happy sciencing!

—Thomas Kardos

National Science Education Standards for Middle School

The goals for school science that underlie the National Science Education Standards are to educate students who are able to

- experience the richness and excitement of knowing about and understanding the natural world;

- use appropriate scientific processes and principles in making personal decisions;

- engage intelligently in public discourse and debate about matters of scientific and technological concern; and

- increase their economic productivity through the use of the knowledge, understanding, and skills of the scientifically literate person in their careers.

These abilities define a scientifically literate society. The standards for content define what the scientifically literate person should know, understand, and be able to do after 13 years of school science. Laboratory science is an important part of high school science, and to that end we have included student labs in this series.

Between grades 5 and 8, students move away from simple observation of the natural world and toward inquiry-based methodology. Mathematics in science becomes an important tool. Below are the major topics students will explore in each subject.

- Earth and Space Science: Structure of the earth system, earth's history, and earth in the solar system

- Biology: Structure and function in living systems, reproduction and heredity, regulation and behavior, populations and ecosystems, and diversity and adaptations of organisms

- Chemistry and Physics: Properties and changes of properties in matter, motions and forces, and transfer of energy

Our series, *Easy Science Demos and Labs*, addresses not only the national standards, but also the underlying concepts that must be understood before the national standards issues can be fully explored. By observing demonstrations and attempting laboratory exercises on their own, students can more fully understand the process of an inquiry-based system. Cross-curricular instruction, especially in mathematics, is possible for many of these labs and demonstrations.

Suggestions for Teachers

1. A • (bullet) denotes a demonstration. Several headings have multiple demonstrations.

2. **Materials:** Provides an accurate list of materials needed. You can make substitutions and changes as you find appropriate.

3. Since many demonstrations will not be clearly visible from the back of the room, you will need to take this into account as part of your classroom management technique. Students need to see the entire procedure, step by step.

4. Some demonstrations require that students make observations over a short period of time. It is important that students observe the changes in progress. One choice is to videotape the event and replay it several times.

5. Some demonstrations can be enhanced by bottom illumination: Place the demonstration on an overhead projector and lower the mirror so that no image is projected overhead.

6. I use a 30-cup coffeepot in lieu of an electric hot plate, pans, and more cumbersome equipment to heat water for student experiments and to perform many demonstrations.

7. As the metric system is the proper unit of measurement in a science class, metric units are used throughout this book. Where practical, we also provide the English equivalent.

8. Just a few demonstrations may appear difficult to set up, for they have many parts. Be patient, follow the listing's steps, and you will really succeed with them.

Safety Procedures

- Follow all local, state, and federal safety procedures. Protect your students and yourself from harm.

- There is no eating or drinking in the lab.

- Be extra careful with broken glass and its disposal. Have appropriate cleanup equipment on hand.

- Wear protective gear like goggles, gloves, lab coats, or smocks when appropriate. Always wear gloves when handling living animals.

- Wear required safety equipment when handling hazardous materials, such as laboratory acids or anything stronger than ordinary vinegar.

- Label all containers and use original container.

- Recognize the value of other living animals in the classroom; take special care to treat animals respectfully. Do not cause pain or harm to living animals during or after an experiment. When weather permits, release invertebrates outdoors after demonstration. We believe it is valuable for young people to interact with living specimens and to share in responsibility for their care and well-being.

- Attend safety classes to be up-to-date on the latest classroom safety procedures. Much new legislation has been adopted in the recent past.

- Have evacuation plans clearly posted, planned, and actually tested.

- Dispose of demonstration materials in a safe way. Obtain your district's guidelines on this matter.

- Have an ABC-rated fire extinguisher on hand at all times. Use a Halon™ gas extinguisher for electronic equipment.

- Learn how to use a fire extinguisher properly.

- Neutralize all acids and bases prior to their safe and approved disposal.

- Have students wash their hands whenever they come into contact with anything that may be remotely harmful to them, even if years later, like lead.

- Conduct demonstrations at a distance so that no one is harmed should anything go wrong.

- Practice your demonstration if it is totally new to you. A few demonstrations do require some prior practice.

Disclaimer: These safety rules are provided only as a guide. They are neither complete nor totally inclusive. Special safety considerations are listed where appropriate immediately following demonstrations and labs. The publisher, editor, and author do not assume any responsibility for actions or consequences in following instructions provided in this demonstration book.

Demos and Labs

When living things act or make a change, it is in response to some change in their environment—a stimulus. A stimulus can be heat, light, sound, electric energy, etc. In general, most animals respond quickly to stimuli, while plants respond slowly and in many different ways. There are always exceptions to this: The Venus flytrap, a plant, responds rapidly, while sea anemones, which are animals, respond more slowly.

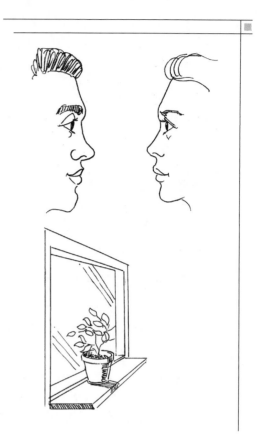

Materials: none

- Ask all students to pair off. They should sit facing each other, about 6 inches apart, and look at each other's eyes. Turn off the room lights. Wait about 15–20 seconds, and then turn the lights on (stimulus). Students should immediately notice the change in the **pupils** of their partner's eyes (response), due to the sudden change in light.

- Take a plant and turn it so that its leaves face away from a window. In a few days, students should notice that the leaves turn toward the widow light. The **stimulus** of sunlight has evoked a **response** in the plant.

Plants actually move during growth in response to certain environmental stimuli. **Auxins,** growth **hormones** produced in the tips of sterns and roots, are responsible for growth movements in plants, called **tropism.**

Sensitivity to gravity is called **gravitropism,** or geotropism. This can be both positive and negative. Positive means moving toward the stimulus, and negative means moving away from the stimulus. In this activity, you will observe the effect of gravity on plant growth. An example can be seen in a plant. The roots grow downward (positive geotropism), while the stems grow upward (negative geotropism).

Materials: corn seeds, empty glass jar with lid, layer of nonabsorbent cotton, paper towels, small amount of clay, water

- Soak the corn seeds in water for a couple of days. Arrange the seeds evenly on the bottom of the jar, with their pointed ends facing inward. Place a layer of nonabsorbent cotton over them, and then fill the jar with paper towels to hold the cotton and the seeds in place. Wet the paper towels thoroughly and place the lid securely on the jar.

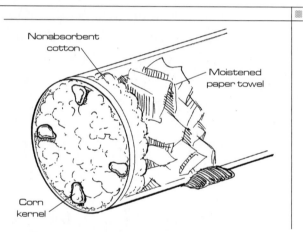

Place the jar on its side, so that one seed points up; you can use a few pieces of clay along the side of the jar to keep it from rolling. Place the jar in a dim light. Make observations through the bottom of the jar over a five-to seven-day period. You will observe the new growth pointing downward.

Phototropism is the movement of plants or animals toward or, in some cases, away from light. In the animal world there are many examples, like a black spider or deep-ocean fish, that move away from light.

Materials: three or four plastic sandwich bags, three or four paper towels, four or five beans per bag (different kinds are fine), water, tacks or staples

If the paper towel is wider than the sandwich bag, cut it so it is slightly narrower than the bag width.

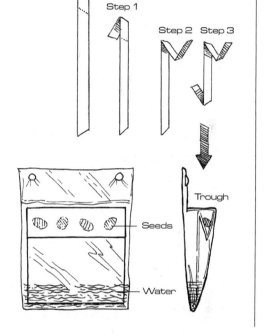

- Follow these steps:

1. Prepare a trough by folding the paper towel one inch from the upper edge.

2. Make another fold an inch below the first one, but in the opposite direction; viewed from the side, the folds should look like the letter N.

3. Fold back the lower end of the paper towel so that the trough assembly will fit inside the bag. Place four or five beans in the trough. Attach the plastic bag to a wall or bulletin board with tacks or staples. Fill the bottom of the bag with about one-half inch of water; water it every few days. If you start the activity on a Thursday or Friday, you will have **germination** by Monday. The plants that grow will have exaggerated stalks pointing toward the room lights or windows. Plant several bags with beans in different locations of your room to demonstrate plant growth toward light.

Many vegetables can grow in midair and even upside down. They will grow in the direction of light sources. This is phototropism. Plants have certain requirements for life, such as light, nutrients, carbon dioxide, etc. This is a great example of how the roots grow toward light in preference to growing toward gravity.

Materials: paring knife, large carrot, small plate, paper towels, wooden toothpick, cotton thread or string, water

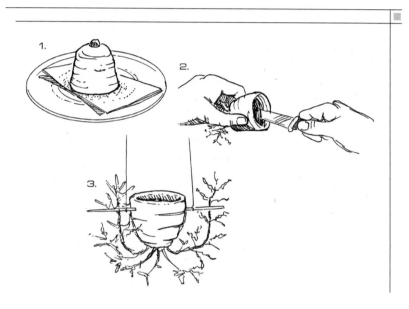

• Use the paring knife to cut a 2- or 3-inch section from the top of the carrot. (Be sure to cut away from your body.) Leave any shoots or stalk attached to the carrot. Place the cut end on the plate over some moist paper towels. Keep the paper moist and keep the plate in sunlight. Once shoots begin to grow out of the carrot, remove the carrot and hollow out its cut side with the paring knife. Poke a toothpick through it and use the toothpick ends and string or thread to hang the plant in a sunny location. Fill the carrot cavity with water and keep it wet at all times. More shoots will sprout from the carrot, and they will all point upward and toward the light. This is due to phototropism.

Materials: two jars with lids, stale bread, water, dirt, crayon

- Place some stale, dry bread in each jar and sprinkle both with dirt. The soil most likely contains mold spores, which may or may not be present in the bread. Sprinkle a few drops of water on one sample and mark this jar with a large crayon mark. Close both jars and place them in a warm place for several days. Notice the growth of **mold** on the bread that was wet. It will have gray or greenish spots.

Stale dry bread with dirt

Stale dry bread with dirt and water

Special Safety Consideration: At the end of the demonstration, dispose of jars without opening them. Mold spores can cause respiratory problems in some people. You may want to ask beforehand if any students have allergies to molds.

As we saw in the demonstration, water is apparently a necessary component for life functions. In the following experiment, we will quantify how much water a particular plant needs to survive and thrive.

Materials: five identical young corn plants, five identical small pots with saucers, graduated cylinder, ruler, water

Procedure:

1. Label the plants 1–5, and measure the height of each plant on the first day, recording it carefully. Make sure all other variables, i.e., sunlight, temperature, air, and nutrients in soil, are identical.

2. Water the plants every day. Give plant #1 no water, plant #2 10 cc (mL) of water, plant #3 25 cc (mL), plant #4 50 cc (mL), and plant #5 100 cc (mL).

3. Measure them every day, and in addition to the height measurements, make a note of their color and apparent health relative to one another. Which is the greenest? Which is turning brown or yellow?

4. At the end of two weeks, write your conclusion.

Conclusion: How much water does this particular plant require? Is there ever a situation when the plant receives too much water? Not enough? Which amount of water seemed to be the correct amount for the plant's health and growth?

Most plants and animals can go for days without food or water, but the vast majority of organisms will die in just minutes without air. Air is important; it is a mixture of gases, and one-fifth of it is oxygen, which is essential for animal life. Most **organisms (aerobic organisms)** get their oxygen from the atmosphere or from dissolved oxygen that is present in water. Strict **anaerobic organisms,** on the other hand, find oxygen toxic, and survive in locations where the oxygen content in the atmosphere or water is extremely low.

Materials: glass tumbler, water

- Fill a glass nearly full of water and let it stand overnight. The next day, students should notice the bubbles on the inside walls of the glass. These are air bubbles, coming from the air that is dissolved in water. The bubbles contain oxygen.

Air bubbles

Oxidation is the rapid process during which fuels combine with oxygen, giving off heat. Humans get their energy from foods, and oxidize mainly sugars. The process, slow in humans, is described as slow oxidation. Humans **inhale** air, which is about 20% oxygen, and **exhale** a reduced amount of oxygen, as well as carbon dioxide, water, and other gases.

Limewater is a colorless liquid. While limewater is available commercially, many schools prepare their own. It is very easy to make. To do so, mix 1.5 g (or .05 oz.) $Ca(OH)_2(s)$ per liter (or 4.25 cups) of water. Stir or shake vigorously and allow the solids to settle overnight. When using limewater, decant carefully to avoid transferring any solids or suspended $Ca(OH)_2(s)$. In the presence of carbon dioxide, this substance turns milky or cloudy. Limewater is used as an indicator; it changes its physical appearance when it reacts with a specific substance. The change is usually in color or turbidity. The change is obvious enough that an untrained person will notice it.

Materials: limewater, glass, drinking straw

Limewater

- Fill the glass about half-full of limewater and start bubbling air through it with the straw. In a short time, the limewater will turn milky white. This shows the presence of carbon dioxide.

Humans inhale oxygen and exhale carbon dioxide, water vapor, and reduced amounts of oxygen. **Bromthymol blue** is an indicator solution that is blue in the presence of bases, pale green when neutral, and yellow(ish) with acids.

Materials: small glass or beaker, drinking straw, 0.1% aqueous concentration of bromthymol blue, water

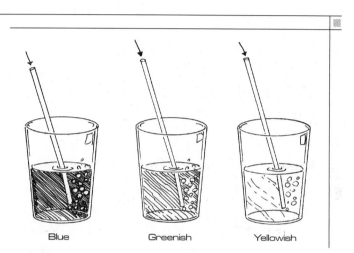

Blue   Greenish   Yellowish

- Place a drop of bromthymol blue in half a glass of water and bubble air through it. The bluish water will appear to become pale—first greenish, then yellowish. The carbon dioxide in your breath mixes with water to form carbonic acid, and bromthymol blue reacts to show the presence of an acid.

Special Safety Consideration: Bromthymol blue can stain clothing and hands, and is a minor skin and eye irritant. It is toxic if taken internally. Take care to wear gloves and goggles, as well as clothing protection, while handling any chemical.

Plants, like animals, respire. Their **respiration** waste includes carbon dioxide and water vapor. In this demonstration, we use limewater to test for carbon dioxide emission by plants. Limewater contains the element calcium, which reacts in the presence of carbon dioxide by turning cloudy. Plants use the process of **photosynthesis,** a photochemical reaction. In this process, plants produce glucose (a form of sugar used as their nutrient) using carbon dioxide and oxygen in the air when activated by the rays of the sun. They release O_2 into the atmosphere. At night, or in the dark, plants release the by-products of respiration, CO_2 and H_2O. Photosynthesis is chemically the opposite of respiration.

Materials: small leafy plant, large jar with lid (big enough to contain the plant), small container (e.g., a baby food jar), limewater. Limewater is easily obtained from school science lab suppliers, or you can easily make your own (see Demonstration 7).

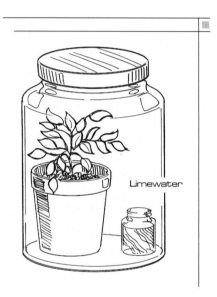

Limewater

- Place the plant inside the larger jar. Fill the small container with limewater and place it inside the jar. Close the jar and store it in a dark place overnight. This prevents photosynthesis and allows only respiration to take place. Notice how the limewater turns cloudy. It is the chemical reaction to the emission of carbon dioxide from the plant during the night.

During the process of respiration, plants in the dark emit carbon dioxide. Plants in the light emit oxygen as a by-product of photosynthesis. Oxygen is one of the three components of fire (along with fuel and kindling temperature), while carbon dioxide puts out fires.

Materials: two sprigs of elodea or other water plant, two test tubes, water, two wide-mouthed jars, matches or lit candle

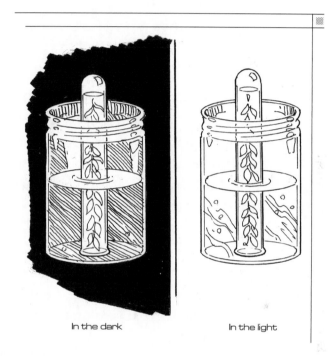

In the dark   In the light

- Place a sprig of elodea into each of two test tubes full of water. Make certain that the two sprigs are of equal length. Place a couple of inches of water in each jar. Place your finger over one test tube, so that you do not lose any water; invert it, and place it inside one jar. Repeat with the other test tube. Now you have two identical setups. Place one in a dark closet, the other in the sun or under a grow lamp. After a couple of days, some gas should have collected at the upper ends of the test tubes. Carefully invert each test tube, keeping it closed with your finger. Holding a burning match or candle near each test tube, let the gas escape. The gas from one test tube will glow bright, while the gas from the other might extinguish the flame.

Special Safety Consideration: Use extreme caution with live flames in the classroom.

Yeast is a tiny fungus that produces carbon dioxide as part of its aerobic metabolism. This makes yeast valuable in the production of bread, wine, and other alcoholic beverages. Yeast metabolizes by fission and **budding.** During fission, the cells divide; during budding, they form a small growth called a bud. During metabolism, yeast produces enzymes that break down sugar. Carbon dioxide is a by-product of this activity, enhancing **fermentation,** which produces ethanol. When carbon dioxide is trapped in dough, it creates the air pockets that make bread rise.

Materials: empty 1-liter soda bottle, balloon, small funnel, several rubber bands or piece of string, empty glass or plastic jar, teaspoon, 5 teaspoons (25 mL) of sugar, 1 teaspoon (5 mL) of yeast (bread or dry yeast work well), water, limewater

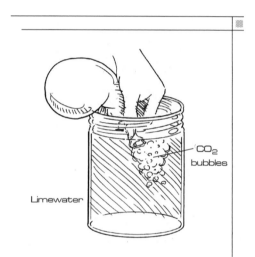

- Follow these steps:

  1. Half-fill the jar with warm water and add the sugar. Stir the sugar until it is dissolved. Pour this solution into the soda bottle.

  2. Wash the jar and add to it the yeast and three teaspoons (15 mL) of water. Stir well, then add to the sugar solution in the bottle.

  3. Attach a deflated balloon to the mouth of the bottle and tie it down tightly with several loops of rubber band or string. Place the bottle in a warm place and observe. The balloon will slowly inflate with carbon dioxide.

  4. Fill the jar with limewater. Carefully pinch the neck of the balloon so that you do not lose any gas, and remove it from the jar. Place the pinched end of the balloon below the surface of the limewater and slowly let the gas bubble out and escape. The limewater, a clear liquid, will turn milky, indicating the presence of carbon dioxide.

The **atmosphere** is a thin layer of gases that surrounds the earth. Among these gases are nitrogen, oxygen, ozone, and carbon dioxide. Since the Industrial Revolution, the amount of carbon dioxide in the atmosphere has been steadily increasing. Infrared rays, the heat energy of the sun, pass easily through the atmosphere. The earth reflects most of the heat back to the sky, but the carbon dioxide in the atmosphere acts as a lid to reflect back and trap some of the heat energy. This has begun a trend in global warming called the **greenhouse effect.** The greenhouse effect has existed since the formation of atmosphere on the planet, and life could not exist on earth without it. However, in modern times, rapid emission of "greenhouse gases," principally carbon dioxide, by human beings is causing a global warming that threatens coastal communities with flooding and may be associated with severe drought. Most nations are starting to address this common problem by trying to reduce their carbon dioxide output. In this activity, you will compare plant growth in an **environment** that closely duplicates this phenomenon with plant growth in a normal environment.

Materials: two seed trays, soil, seeds for flowers, water, plastic food wrap, two thermometers, large rubber band, wooden skewers or ice-cream sticks, pencil, notebook

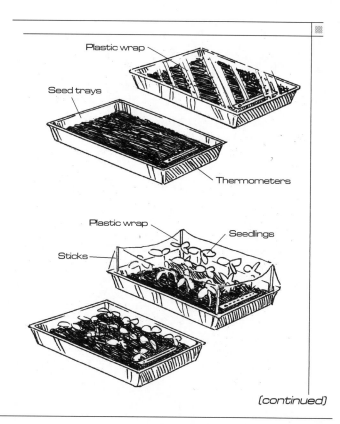

- In cold climates, do this demonstration in the spring; in warm climates, any season will do. Fill the seed trays with soil, and plant several kinds of flower seeds in them. Make certain that the soil is moist. Place a thermometer in each tray. Cover one tray with plastic wrap, securing it by placing the large rubber band around

(continued)

the tray. Place both trays in an outside location. Shelter them from rain but expose them to sunlight. As the seedlings grow taller, insert the ice-cream sticks in the corners of the wrapped tray to keep the plastic from touching the plants, then replace the plastic wrap. Weekly observe and record the temperatures of both trays. You will notice that the closed environment is warmer than the other tray.

Special Safety Consideration: If using mercury thermometers, use extreme caution. If breakage occurs, dispose of mercury properly, according to hazardous material standards.

Plants have tiny openings called **stomata** on the undersides of their leaves. The function of the stomata is to let out excess gases, such as oxygen, and to take in carbon dioxide. If the stomata are blocked, the leaf is unable to take in carbon dioxide and dies.

Materials: petroleum jelly, leafy potted plant (geraniums work well)

- Keep the plant in the dark for several days. This will reduce the plant's metabolism to a minimum. Then select two leaves and coat their upper surfaces heavily with petroleum jelly. Select two more leaves and heavily coat their undersurfaces with petroleum jelly.

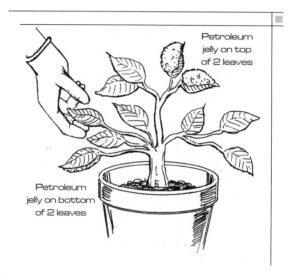

Petroleum jelly on top of 2 leaves

Petroleum jelly on bottom of 2 leaves

- Place the plant in a sunny window for a week and observe. The light from the window will increase the plant's metabolism and photo-synthesis. The leaves coated on the underside will die as time goes on.

Plants absorb water from the earth. Water moves up from the roots, through vessels and **tracheids,** to reach the leaves. Most of the water in the leaves, about 90%, is lost through the stomata. This water loss through the stomata is called **transpiration.** Since large trees can lose as much as 6800 kg (14,900 lb) of water in 12 hours, this transpiration can affect the local weather (microclimate) in terms of humidity and temperature. A plant that runs out of water in the soil will transpire itself to death.

Materials: leafy plant, sandwich bag, sealing tape

- Place the sandwich bag over one leaf of the plant. Use tape to seal the bag tightly around the stem. Place the plant in sunlight for several hours. Observe the interior of the bag for cloudiness, as water collects on its inner surface. After several hours in sunlight, the bag will show water condensation and will be slightly cloudy.

How can you measure water loss in plants? All plants, as we now know, lose water through transpiration. Transpiration occurs when water travels through a plant to its leaves and escapes through pores as water vapor.

Materials: graduated cylinder, potted plant, reclosable plastic bag, water

Procedure:

1. Allow the soil of a small potted plant to dry out so that the plant begins to wilt a tiny bit.

2. Using the graduated cylinder, water the plant just until water starts dripping out of the bottom.

3. Allow the water to drip back into the cylinder.

4. When it stops dripping, subtract what remains in the graduated cylinder from the amount originally in it. This is the record of the amount of water currently in the plant's soil.

5. Place the plant in the reclosable bag and seal it.

6. Place the plant in the sunlight.

7. After three days, there should be a good deal of water on the inside of the bag, and the plant will be drying out.

8. Remove the plant, reseal the bag, and place it in a refrigerator for an hour. Then pour out as much water as you can into a graduated cylinder. This is the amount of water the plant has transpired during that period.

Conclusion: If the potted plant had not been placed in a bag, where would the water have gone? Would the results have been different if you had placed the plant in a shady location? Why or why not?

Like a greenhouse, a terrarium is an environment that encourages the greenhouse effect. Plants in a terrarium create their own **ecosystem,** complete with water cycling and use of nutrients in dead and decomposing material. The typical terrarium has a number of different plant species, all adapted to a particular ecological zone, such as a rainforest or desert environment. In our experiment, we will look at the role of this **diversity** in the small world of a terrarium. In a rainforest, there are many different species competing for different nutrients, light, and water requirements. In a cornfield, on the other hand, there is a large number of individuals of the same species competing for the same nutrients, light, and water requirements.

Materials: two large jars (industrial-sized pickle or mayonnaise), some small pebbles, activated charcoal, potting soil, a variety of small plants with different light requirements for one jar and a single variety of small plant for the other, water, and a small watering can or small beaker

Procedure:

1. Rinse the pebbles thoroughly. On the bottom of your clean jars, layer 1 or 2 cm of pebbles, topped by a centimeter of activated charcoal. Cover these with 8 cm of potting soil. Be sure the soil doesn't clump.

2. In one jar, plant your "rainforest," a variety of small plants. Plant each by scooping out a couple of centimeters of soil, placing the plant in the depression, and covering it well. In the other, do the same for your single-variety "cornfield." Water both lightly—about 50 mL each—and cover the terrariums with the jar lids. (Be sure you don't accidentally leave a lot of soil or mud on the glass, or it will be difficult to see what is going on inside the jars.)

3. Do not open the lids, except to water once every other week.

4. After a month, look at your terrariums. Which appears to be the healthier environment? How can you tell?

Conclusion: What is the role of diversity in an ecosystem? If diversity is impossible, as in a cornfield, what kind of intervention is necessary for healthy growth?

Plant transpiration can be verified using a chemical **indicator,** a strip of **cobalt chloride.** While previous tests show the presence of a liquid (which we assume is water), this test verifies that the liquid is actually water.

Materials: small leafy plant, large jar with lid, cobalt chloride indicator paper (available at most scientific supply houses)

Cobalt
chloride
paper

Cobalt chloride is an indicator that changes color in the presence of water vapor. To see it work as a first demonstration, you can breathe on a strip and watch the color change from blue to pink.

- Place the plant and a strip of cobalt chloride in the jar. Cover the jar and set it in a dark place overnight. The paper will turn from cobalt blue to pink. The pink indicates the presence of water.

Special Safety Consideration: Cobalt chloride, as an indicator strip, is relatively harmless; however, the chemical in its solid state is a known human carcinogen. It may also cause lung problems that may not be apparent immediately. Be sure to collect and dispose of indicator strips immediately after use. Use gloves every time you use cobalt chloride, since perspiration from your bare hand will change the color of the indicator strip.

The shape of a leaf determines how quickly water will transpire from it. On a broad leaf with a larger surface area, water will transpire faster than on a narrow one with a smaller surface area. Desert plants have leaves that are thick and round with a greatly reduced surface area. They are also waxy, which inhibits evaporation. This explains why desert plants have a very slow rate of transpiration. Other factors that control transpiration rates include the opening and closing of stomata and the humidity level in the surrounding air.

Materials: four paper towels, wax paper, paper clips, water

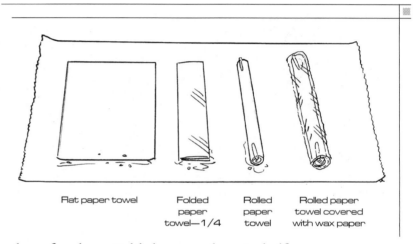

Flat paper towel   Folded paper towel—1/4   Rolled paper towel   Rolled paper towel covered with wax paper

- Dampen all paper towels equally so that they are wet but do not drip. Lay one towel flat on a piece of wax paper that is two or three feet long. Fold the next sheet in half, then in half again, and place it on the wax paper next to the first towel but not touching it. Roll up the third and fasten the ends with two paper clips, then place it on the wax paper. Finally, place the last towel on a piece of wax paper the same size as the paper towel and roll the two sheets up together, fastening the ends with paper clips. Place it on the wax paper with the other sheets. Place the supporting wax paper with the wet sheets in direct sunlight. A day later, observe the wetness of the sheets. The flat sheet should be dry; the folded sheet might have a damp spot or two near its bottom; the rolled-up sheet will have several damp spots; the wax-covered sheet will still be damp all over.

Leaves and stems are shaped by the woody vessels that conduct water through the plant. These vessels can be compared to the human system of arteries and veins. Holding up a leaf to the light will let you see this woody structure. To study it more closely, you can isolate the structure and mount it on cardboard. Obtain a variety of leaves from a local park or your garden. Use a tree guidebook to identify the leaves.

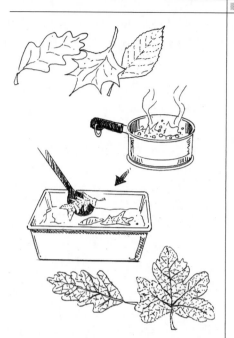

Materials: water, leaves, washing soda (sodium carbonate decahydrate), saucepan, plastic tub, slotted spoon, household bleach (sodium hypochlorite), paper towels, cardboard, newspapers, rubber gloves, hot plate

- Fill the saucepan about three-quarters full of water. Add 1 teaspoon (5 mL) of washing soda and bring it to a boil. Add the leaves and boil them for one hour. If you have too many leaves, do it in two batches. In the plastic tub, make a solution of water and household bleach, six parts water to one part bleach. Wear rubber gloves and be careful not to spill either the bleach or the bleach solution. Using the slotted spoon, carefully place the boiled leaves in the bleach solution. By the next morning, everything but the woody structures of leaves will have floated away. Remove the leaves and arrange them on paper towels (on top of several layers of newspaper) to dry. When the leaf structures are dry, they are ready for mounting on cardboard and labeling.

Special Safety Consideration: Protect your clothes and work area from the bleach. Immediately wash anything that gets splashed with bleach, bleach solution, or the washing soda solution. If your skin comes in contact with washing soda, bleach, or bleach solution, rinse the affected area under running water. Wear protective garments, gloves, and goggles.

Seeds germinate and grow under specific conditions. Different varieties of seeds may do best under different conditions. This activity will address four variables: temperature, light, water, and air. It is important to note that when you test for one variable, the other three must be kept unchanged. Corn or beans are good choices for this demonstration, as they germinate quickly. You will use two dishes: one to experiment with, the other as a control. The results will demonstrate which variables are necessary for germination.

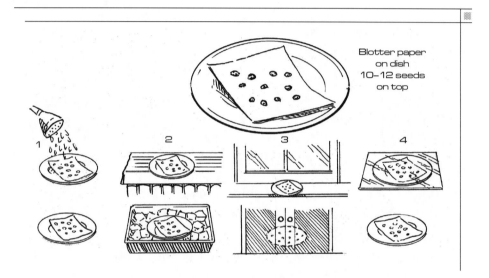

Blotter paper
on dish
10–12 seeds
on top

Materials: eight small dishes, corn or bean seeds, wax pencil or waterproof felt pen, baking pan, ice cubes, blotting paper or cotton, small piece of glass (as from picture frame), water

- Place a piece of blotter paper or a thin layer of cotton in each of the eight small dishes. Number each one. Place 10 or 12 seeds of the same variety on top of the blotter or cotton in every dish, then expose each dish to the conditions described in each pair of experiments on the following page:

(continued)

Variables	Constants
1. Water: Water one group of seeds, but do not water the other.	Temperature, air, light
2. Temperature: Keep one group of seedlings in a warm area; keep the other over a pan of ice cubes.	Air, light, water
3. Light: Keep one group of seedlings in a sunny window; keep the other in a dark closet.	Air, water, temperature
4. Air: Keep one group of seedlings under glass; expose the other to air.	Temperature, light, water

After several days, observe the results in each dish. Plants in a warm dish will grow better than plants on ice. Seeds in the dark may germinate as well or better than those in the light, as light is not needed for germination. Seeds left without either air or water will not germinate.

J. Weston Walch, Publisher
Easy Science Demos & Labs: Life Science

Plant **chloroplasts** contain **chlorophyll.** This gives plants their characteristic green color. Metabolism in a plant takes place when the sun's energy reaches the chloroplasts to produce sugar from carbon dioxide and water. This process is called photosynthesis. Without chlorophyll, the process cannot take place. To study chlorophyll, scientists separate the chlorophyll from its source leaves. This process, called chromatography, is performed in this activity.

Materials: beakers, isopropyl alcohol, filter paper, several leaves from different plants

- Cut several strips of filter paper about 1 inch wide and 6 inches long. Chop up leaves, place them in separate beakers, and soak them in alcohol for about an hour ahead of this demonstration. About 3 inches from the bottom of each strip, place a dot of the greenish liquid from each of the leaf beakers. Tape the strips to the inside edge of a clean beaker and pour about half an inch of alcohol into the beaker. Look at the strips 5 minutes later. Notice how this process separates the elements of the chlorophyll.

The pressure that exists inside a **cell** is called the **turgor pressure.** As water diffuses into a cell, the pressure on the **cell membrane** or cell wall increases, and we say the turgor pressure also increases. If turgor pressure is maintained, plant tissue, having a cell wall in addition to a cell membrane, is usually more rigid than animal tissue. When the turgor pressure decreases, the plant wilts.

In this experiment, we will expose plant cells to a **hypertonic solution.** That is, the concentration of dissolved substances outside the cell (in this case, salt) is greater than the concentration inside the cell. If a cell is placed in a hypertonic solution, **osmosis** will cause water to leave the cell.

Materials: small water plant (elodea works well), 3% salt solution, microscope slide, coverslip, paper towels, microscope, forceps

Procedure:

1. Use forceps to take a leaf from the tip of a water-based plant, such as elodea, and place it on a slide on a drop of water. Cover it with a coverslip.

2. Observe the slide under a microscope. Find the leaf under low power and then observe it under high power. Draw a picture of a cell and its contents.

3. Place a drop of 3% salt solution on the slide near one end of the coverslip. Use the paper towel on the other side of the coverslip to draw out the plain water and draw the saline solution under the slip.

4. Observe a cell and then sketch it.

Conclusion: What happened to the cells of the plant when they were exposed to the saline solution? Explain in terms of diffusion and osmosis what took place.

Humans inhale air, which is a mixture of oxygen and other gases—mostly nitrogen. We exhale water vapor, nitrogen, a smaller amount of oxygen, carbon dioxide, and other trace gases.

Materials: small mirror

- Breathe on a mirror to show that you are producing water vapor. Your breath will condense, leaving a haze of moisture on the glass.

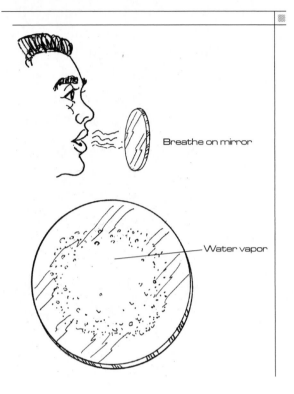

Breathe on mirror

Water vapor

Animals inhale oxygen and exhale carbon dioxide. Plants and other chlorophyll producers in turn absorb the carbon dioxide and generate oxygen, a mutually life-sustaining process. This is part of the reason that in fish tanks and marine habitats, chlorophyll producers are needed to oxygenate the water for fish, while fish produce the carbon dioxide needed for plants to manufacture food.

Materials: goldfish, glass or jar, limewater

Limewater

- Place the goldfish in a glass of limewater and let students observe how, in a short time, the limewater turns milky white, indicating the presence of carbon dioxide. Rinse the fish and place it promptly back into its own aquarium so that it is not harmed.

Plants and other chlorophyll producers use carbon dioxide in their metabolism and produce oxygen as a by-product.

Materials: jar with lid, soda water, elodea or other water plant, test tube, florist's clay, nail, hammer, small block of wood, matches, splint

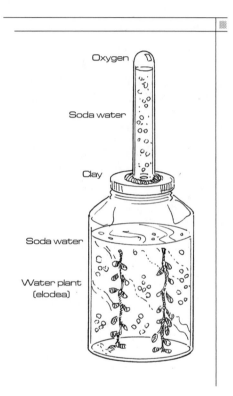

- Set the water plant in a jar nearly full of water and set the jar in sunlight until bubbles appear on the plant. These are oxygen bubbles. During photosynthesis, a plant's food-manufacturing process, plants use sunlight and water to produce simple sugars and oxygen. Chlorophyll, which produces the green color in leaves and stems, absorbs the light energy for the plant.

- Place a water plant in a jar containing soda water, which is rich in carbon dioxide. Allow the soda water to sit for at least half an hour before inserting the plant. Use the nail to pierce a hole in the jar lid: Place a small wooden block under the lid, and hit the nail with the hammer. Place the lid on the jar carefully; place an inverted test tube filled with soda water over the hole in the lid. Seal all contact edges with florist's clay. Place the jar in sunlight. After the plant begins to metabolize, some water in the test tube will be replaced with oxygen. You can test for this oxygen by removing the test tube and inserting into it a glowing wooden splint. Light the splint, blow off the flame, and while it is still glowing, insert it into the test tube. The splint will relight in the presence of oxygen.

It is exciting for students to observe the **circulation** system of live
animals without harming them.

Materials: small glass dish, cotton
wrapping, water, goldfish, microscope set at a low power

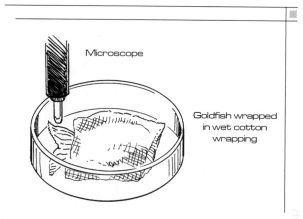

Microscope

Goldfish wrapped
in wet cotton
wrapping

• Wrap a goldfish in wet cotton
wrapping, leaving the tail
exposed, and place it in the dish.
Focus the microscope on the
tail of the fish (outside the
wrappings) and observe the many
capillaries (a type of **blood
vessel**). You will also notice red corpuscles moving in various
directions. Those moving toward the tail are in **arteries,** while
those moving in the opposite direction travel through veins.
(Note that microscope images are reversed, so that something
that appears to be traveling toward the tail is actually traveling
away from it.) Be sure to return the goldfish to its aquarium
promptly, so that it is not harmed.

Fish have been swimming now for 500 million years, so it is not surprising that they have become very good at it. There are three basic ways that fish move through the water. Some, like eels, swim in an S-shaped pattern, alternatively tightening muscles on either side of the body. Others, like mackerels, flex the back part of the body as well as the tail, but leave the head and front of the body rigid. Still others, like the tuna, keep the entire body rigid, moving only the tail. Fish like this move faster than the others.

Materials: community aquarium of different fish

Procedure:

1. Observe several fish from the top of the aquarium, so you can watch their body movements in full. Determine which pattern they follow.

2. Drop in a little fish food, and see if any of their swimming patterns change.

Conclusion: What might be the survival value for the different swimming patterns? How might the swimming patterns affect feeding and mating habits?

Water and other molecules tend to stick to their own kind. This is called **cohesion.** When molecules stick to other kinds of molecules, the process is called **adhesion.** When water or other substances enter very fine openings, they rise through a combination of cohesion and adhesion. This process is called **capillary action.** It describes how leaves of a very large tree can grow by drawing water from the tree's roots, through the stems, and up through the trunk and the branches, until it finally reaches the foliage. Capillary action is not, in and of itself, responsible for water rising through a plant. In addition to capillary action, transpiration serves to pull the water up, as water is released from the stomata of the plant, leaving a void.

Materials: small jar, water, red food coloring, celery stalk

- Half-fill the jar with a deep-red solution of food coloring and water. Place a celery stalk in the jar. Let the setup stand for 15–20 minutes. Cut the celery stem above the waterline and observe the red color in its vessels.

Celery stalk

Red food coloring in water

Red color in tubes

Most animal cells are composed of a nucleus, cytoplasm, and organelles that have discrete functions. Unlike plant cells, animal cells have no hard cell wall; rather, they possess a permeable membrane through which nutrients and waste products pass. Animal cells have different shapes based on their function; there is a world of difference between a **white corpuscle,** a nucleus-free **red blood cell,** and a star-shaped nerve cell. For simplicity's sake, we will construct a prototypical animal cell with a "nucleus," "organelles," and "cytoplasm" as a demonstration model, but students should be informed that cells have different appearances based on their function in the animal's body.

Materials: small plastic bowl, water, package of gelatin, bean, tiny seashells, plate

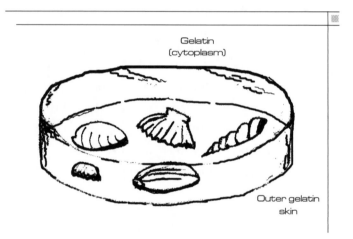

Gelatin (cytoplasm)

Outer gelatin skin

- Prepare the gelatin according to the instructions on the package. Pour the hot gelatin into the plastic bowl and let it cool. When the gelatin is nearly hardened, place the bean in the center of the gelatin and the little seashells in other locations in the gelatin. Let the gelatin harden. Flip the bowl over and lay the gelatin on a plate or other flat surface. Now you have a model of an animal cell with the nucleus (bean), organelles (seashells), the cytoplasm (gelatin), and the cell membrane (outer skin of gelatin).

Robert Hooke discovered and named cells when he first looked at cork under his early microscope. Cork is a dead, protective layer of plant cells that is found on the trunk of some species of trees, especially the cork oak.

In this experiment, we will be looking at cork cells under the microscope and comparing them to the animal-cell model we constructed in the last demonstration.

Materials: cork, microscope, microscope slide and coverslip, stain, sectioning tool or razor blade

Procedure:

1. Shave a paper-thin slice of cork from the stopper with the razor blade or sectioning tool.

2. Place the cork on a drop of water or glycerin; on a microscopic slide, stain it to see the details better and cover with a coverslip.

3. Observe the cells at the edge of the cork at both low and high power under the microscope. Make a drawing of the cells at each power.

Conclusion: What shape are cork cells? What is inside the cork cells? What function does the cork provide for the tree? How are plant cells different from animal cells?

Strong Safety Warning: Use caution while handling the razor blade or sectioning tool. Sharp objects can cause injury.

Starch is a white, odorless, tasteless, granular or powdery complex **carbohydrate** $(C_6H_{10}O_5)_x$ that is the chief storage form of carbohydrates in plants. It is an important foodstuff, and is also used in pharmacy and medicine. Carbohydrates are foods that contain the elements carbon, hydrogen, and oxygen. Examples of carbohydrates include starch, sugar, potatoes, vegetables, fruit, cereals, and legumes.

Materials: **Lugol's solution,** bread, cracker, potato slice, and other foods containing little or no starch, such as a slice of ham, hamburger, etc. Lugol's solution is a brown liquid containing iodine.

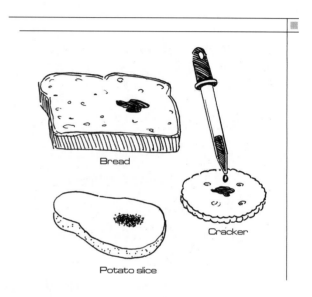

Bread

Potato slice

Cracker

- Add a drop of Lugol's solution to a slice of bread, a cracker, or a slice of potato, and a slice of ham or a bit of ground hamburger. If the Lugol's solution turns blue-black, then the substance contains starch.

Special Safety Consideration: Lugol's solution is toxic if taken internally. Wear gloves and goggles, and follow all safe disposal procedures.

Simple sugars are formed in plants. They are the major product of the **carbon cycle,** which utilizes carbon dioxide, sunlight, and water to produce sugars and oxygen. There are many types of sugar, from fructose, which is formed in fruit, to sucrose, which is the end product of cane and beet sugar, to glucose, to which all sugars we eat convert in our blood. There are several other sugars. Sugar is an important part of our diet, since it provides the human body with energy. It isn't necessary to eat refined white sugar, because sugars are found in many natural foods, such as fruits and vegetables.

Benedict's solution is a bluish liquid. When it is heated in the presence of sugar, it changes in color from blue to green, yellow, orange, or red.

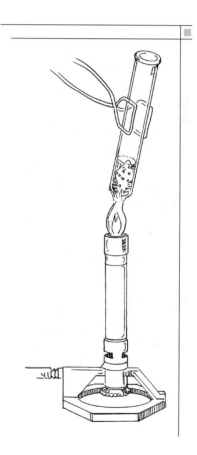

Materials: Bunsen burner, Benedict's solution or Clinitest™, test tube, forceps, foods that contain sugar (fruit, cake, etc.), as well as foods with little or no sugar in them, such as artificially sweetened candy and drinks

- Place the substance you wish to test in a test tube and add a few drops of Benedict's solution. Using forceps, hold the test tube over the Bunsen burner. If the solution changes color as it is heated, the substance contains sugar.

- Alternatively, buy a package of Clinitest at a pharmacy or drugstore. (You may have to order it in advance.) Follow the directions on the package and test other foods for sugar. With Clinitest, you do not have to heat foods to test for sugar. As with most other indicators, you match colors.

Special Safety Consideration: Benedict's solution is toxic and can spatter when heated. Wear gloves and safety goggles. While heating, tilt test tube away from students.

Most foods contain **minerals.** When burned, they leave a residue—a gray or whitish ash. The ash indicates the presence of minerals. If no ash is left, the food contained no minerals. The ashes show the presence of minerals but do not show which minerals. Other tests are available to test for specific minerals.

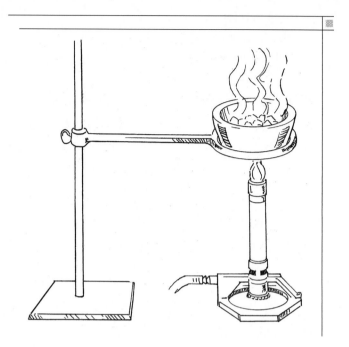

Materials: small crucible, crucible support, ring with stand, Bunsen burner, food to be tested for minerals (cheese will leave calcium, and dark leafy vegetables will leave iron, for instance)

- Place food to be tested in the crucible and heat until it either burns off or turns into ash.

Note of Interest: Since matter is conserved, the rest of the food item continues to exist in a gaseous state.

Materials: brown paper bag, foods to be tested for **fat.** Use foods that are high in fat as well as those known to be low in fat to demonstrate a difference, such as cheese versus an apple. It will be easier to work with fruits or vegetables that are past their prime; or cut the fruit and allow the juice to soak in.

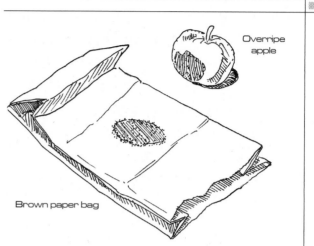

Overripe apple

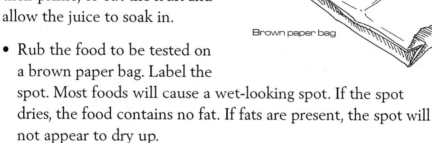

Brown paper bag

• Rub the food to be tested on a brown paper bag. Label the spot. Most foods will cause a wet-looking spot. If the spot dries, the food contains no fat. If fats are present, the spot will not appear to dry up.

Proteins belong to a group of complex organic (carbon-containing) compounds that are an important part of cell structure. Proteins are an important component of the human diet, since they are primarily responsible for growth and repair of body tissues.

Materials: copper sulfate, lime (garden lime), water, plate, two eyedroppers, two glass jars with covers, foods to be tested for protein— meat, peanut butter, cheese, slice of hard-boiled egg, carrot, piece of apple

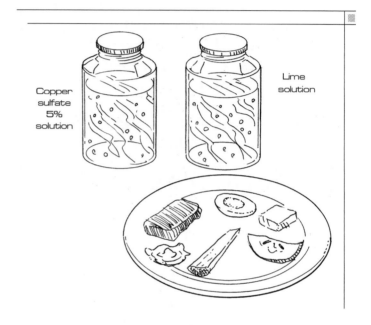

- Prepare in one jar a 5% solution of copper sulfate: 5 mL copper sulfate to 95 mL water. In the other jar, prepare a lime solution: 225 mL of water to 15 mL of lime.

Place the foods to be tested on a plate. Add one eydropperful of both solutions to each food to be tested. If a violet color appears, the food contains protein. A darker violet indicates more protein than a lighter violet.

Proteins are foods that contain the element nitrogen. Examples include meat, fish, fowl, eggs, and cheeses.

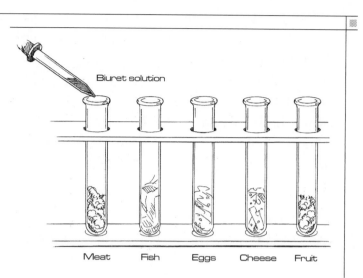

Biuret solution

Meat   Fish   Eggs   Cheese   Fruit

Materials: **biuret solution,** test tube, test tube rack, foods to be tested for protein

- Biuret solution is an indicator used to test for food proteins. Add food to the test tube, then add a dropperful of biuret solution. If the solution changes from light blue to purple, then protein is present. The darker the purple, the more protein there is.

Vitamin C is an essential **vitamin** that helps living cells to grow and reproduce. The human body needs a good supply of vitamin C if it is to function properly. A vitamin C deficiency results in scurvy, a serious illness. It starts with bleeding gums and tooth loss, and was the dreaded scourge of sailors in centuries past. Since vitamin C is water soluble and does not accumulate in the human body, we need a fresh supply of it on a daily basis. Citrus fruits and bell peppers are good sources of vitamin C.

Materials: **indophenol,** test tube, stopper, eyedropper, food to be tested, including foods rich and deficient in vitamin C. C-rich foods include citrus fruits, peppers, and tomatoes. C-deficient foods include most protein foods, sugar, and starches.

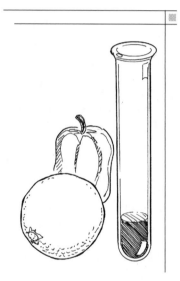

- Indophenol is an indicator used to test for vitamin C. Place about 1 inch of indophenol in a test tube. Add a small amount of food, a small bit at a time, to the indophenol. Replace the stopper and shake well after each small amount of food is added. Indophenol is light blue and will turn colorless in the presence of vitamin C.

A typical human being can survive for months without food but will die in three days without water. Our bodies are almost 70% water, and so are the bodies and structures of every living thing on earth. Because we eat previously living things—animals and plants—we consume with our proteins, sugars, vitamins, minerals, and starches an enormous quantity of water. Depending on how they are processed, some foods on our tables will contain more water than others. This test will demonstrate how much water is present in various foods.

Materials: test tubes, forceps, food samples (cracker, cheese, slice of fruit), Bunsen burner

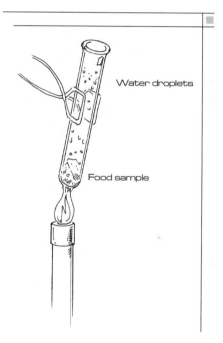

Water droplets

Food sample

- Place a small sample of food in a dry test tube and heat the test tube gently. If droplets of water form on the inside of the test tube, there is water in the food. If no water droplets form inside the test tube, the food does not contain water.

Salt, or sodium chloride, is a naturally occurring compound found everywhere on the planet. Salt has been used since ancient times as a **preservative,** since it dries out water in foods, thus slowing the decay of the food. At certain times, it was highly prized and extremely valuable. In modern times, salt is used as a condiment and heightens the flavor of natural foods.

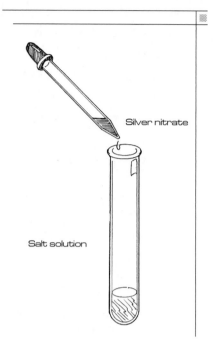

Silver nitrate

Salt solution

Materials: test tube, silver nitrate, 1:10 salt and water solution, eyedropper

- Drop a few milliliters of salt solution in the test tube. Add a few drops of silver nitrate and observe the solid that is formed. The solid indicates the presence of salt.

In soil or water, or exposed to the air, materials decompose through biological means, **biodegradation.** Biodegradation relies on decomposers, such as fungi, **bacteria,** annelids and insects, as well as a variety of other flora and fauna. These are extremely important to the ecosystem, since decomposition could not take place without them. Many manufactured chemicals used in farming, food preservation, and other everyday applications prevent this natural breakdown. A red maraschino cherry, for instance, will not easily decompose, as it has been treated with preservatives that inhibit biodegradation.

Materials: two plastic milk carton bottoms (gallon size), soil, maraschino cherry, piece of bread or fruit, water, oven

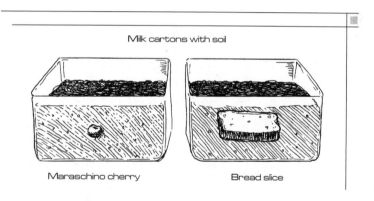

Milk cartons with soil

Maraschino cherry   Bread slice

- Half-fill the bottoms of both milk cartons with dirt. Place a maraschino cherry in one, a piece of bread or fruit in the other. Cover both with soil. Keep the soil moist by watering regularly. Keep a log of watering dates and depth of samples. At the end of one month, look for what you buried. The cherry will have changed color but will be intact, while the bread or fruit will have decomposed. Write comments on the status of the samples when unearthed.

- Set the experiment up again, but this time, bake the soil that will go into one of the containers for an hour at 450°F. Choose two slices of bread from the same package and bury both, taking care not to touch them with your hands. Have students note that the bread in the unsterilized soil is decomposing faster.

Decaying and decayed biological material releases nutrients that combine with topsoil to form humus. The amount of humus in soil depends on the rate of decay, which is based on a number of factors, including pH balance in the soil, temperature, the amount of moisture, and other things. The amount of humus is also based on the rate of absorption of nutrients by plants in the vicinity. Since rainforest plants absorb most of the nutrients in the soil, there is relatively little humus there, while temperate forests have a greater amount.

Materials: plastic cups; clean, washed sand; soil from the area around your school or home; eyedropper; and 2% hydrogen peroxide solution

Procedure:

1. Place 15 mL of washed sand in a plastic cup.

2. Using an eyedropper, drop peroxide into the cup. Bubbles will form as the hydrogen peroxide breaks down any remaining organic material in the sand.

3. Record the number of drops you used until the bubbles stopped forming.

4. Repeat the process for the soil sample from your local area.

Conclusion: Did your soil contain more or less humus than the clean sand sample? Why do you think you saw the result you saw? Could you make a prediction about the amount of humus in the soil in any other **biome,** for instance, the African savannah, or the Mississippi River bank?

Vitamins, nutrients found mainly in plants or in plant-eating animals, affect human health. Here is a brief summary of key **deficiency diseases** caused by vitamin deficiency in the **diet:**

Vitamin	Deficiency Diseases
A	Hard, dry skin; night blindness
B_1	Beriberi
B complex	Neurological disorders
C	**Scurvy**
D	**Rickets**
E	Severe circulatory problems
K	Poor or no blood clotting

- Read to students the labels of a bottle of a multivitamin supplement.

- Read to students the labels on several food packages.

- Have students research the various vitamin-deficiency diseases and have them make brief reports.

Minerals are chemicals that are essential for life. A shortage of minerals in the human diet causes severe problems. A deficiency of iron causes **anemia.** A deficiency of iodine causes the thyroid gland to swell, a condition called **goiter.** Iodine is added to salt in the United States.

Here is a summary of a few minerals and their uses:

Minerals	Key Uses
Sodium	Keeps muscles and nerves healthy
Potassium	Keeps muscles and nerves healthy
Calcium	Building block for teeth and bones
Phosphorus	Keeps teeth, bones, and brain healthy
Iodine	Controls human metabolism (oxidation)
Trace metals (zinc, etc.)	Maintains healthy body functions
Iron	Helps build red blood cells and prevents anemia

- Read to students the labels of several vitamin-plus-mineral bottles.

- Read to students the labels on several food packages.

- Have students research minerals and their deficiency-related problems and have them make brief reports.

Iron is a key mineral added to the human diet to assist the body with the manufacture of red blood cells. According to research, women need more iron than men.

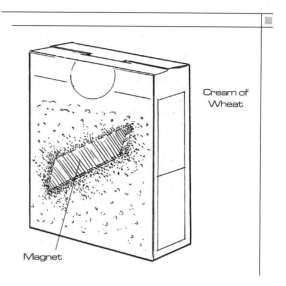

Cream of Wheat

Magnet

Materials: box of Cream of Wheat™, magnet

• Place the magnet inside the box of Cream of Wheat, close the box, and shake it for a couple of minutes. Have your students observe the iron that collects on the magnet.

The digestive tract is made of many tissues, and among these is muscle tissue. **Peristalsis** is the muscular action in the digestive tract that moves the food along its way.

This short demonstration will illustrate how muscle contractions move food along through the digestive tract.

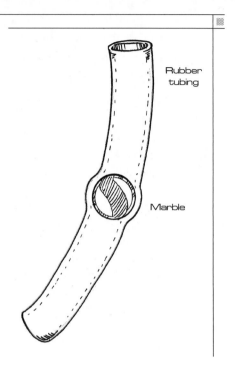

Rubber tubing

Marble

Materials: rubber tubing, marble

- Find a length of rubber tubing that is not wider than a marble. Place the marble inside the tube and move it along by squeezing the tube.

Digestion in most mammals occurs in two discrete steps. Food is masticated (chewed) in the mouth, and it is also exposed to acids in the mouth and one or more stomach acids before passing along into the intestines for absorption of nutrients into the bloodstream. (Some mammals repeat the mastication step one or more times—we call that "chewing the cud.") The first method of digestion is purely mechanical, while the second is chemical.

Materials: small chopping block, celery or apple, sharp knife, two test tubes, sugar cubes, granulated sugar, water

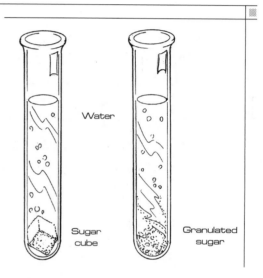

- Chop apple or celery into fine pieces to simulate the action of teeth in the mouth.

- Fill both test tubes about three-quarters full of water. Add a sugar cube to one and $\frac{1}{2}$ teaspoon (2.5 mL) granulated sugar to the other. Shake both for a few seconds and observe how much longer it takes to dissolve the sugar cube. Breaking food down mechanically allows it to be absorbed more quickly into the **digestive system.**

J. Weston Walch, Publisher

Chemical digestion breaks food molecules into smaller ones, to allow the nutrients to be absorbed by the body. **Saliva** in the mouth lubricates and wets the food to allow it to move through the digestive tract; it also provides **enzymes,** chemicals made by the body to help break down the food. Saliva enzymes digest starch and change it into sugar.

Materials: piece of bread or cracker for each student

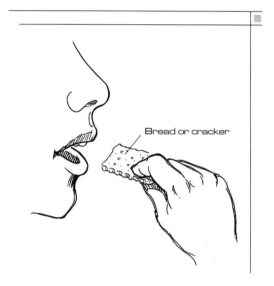

Bread or cracker

- Give each student either a piece of bread or a cracker to chew. Have them chew for about two or three minutes without swallowing. The bread or cracker will taste sweeter. This is a good example of chemical digestion. (See the following demonstration for a lab test of this process.)

Demonstration 41, Chemical Digestion, showed informally that enzymes in saliva convert starch into sugar. This lab gives a formal demonstration of the same thing.

Materials: four test tubes, stoppers, starch, water, Benedict's solution, Lugol's solution (iodine), Bunsen burner, forceps, saliva

- Place a small amount of starch in each of the four test tubes. Assign each tube a number from 1–4. Fill each one-quarter full of water, and shake to mix the contents.

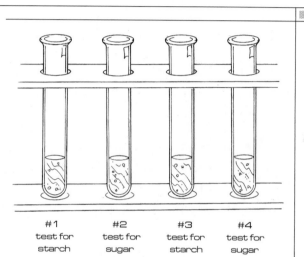

#1
test for
starch

#2
test for
sugar

#3
test for
starch

#4
test for
sugar

1. Use Lugol's solution to test #1 for starch as you did in Demonstration 26. It will confirm that starch is indeed present.

2. Use Benedict's solution to test #2 for sugar. (For details, see Demonstration 27, Testing for Sugar in Foods.) The test will confirm that #2 does not contain sugar. Test tubes #1 and #2 form your control group.

3. Now place some saliva in the remaining two test tubes. Stopper them and shake them well. Use Lugol's solution to test #3 for starch. The result will be positive—some starch is still present.

4. Test #4 for sugar. The result will show that it contains sugar. The enzymes in the saliva have changed some of the starch into sugar.

While the saliva in the mouth contains one enzyme that changes starch into sugar, other nutrients must be separated too. The stomach produces gastric juices that lubricate the food so that it can move along. **Gastric juices** contain the enzymes **rennin** and **pepsin.** In addition, they contain **hydrochloric acid** (HCl) and water. The acid kills bacteria and dissolves minerals.

In this demonstration, we will replace HCl with simple white vinegar to illustrate how acids affect foods in the stomach.

Materials: chicken bone, piece of string, beaker, vinegar, safety gloves, soap, and water

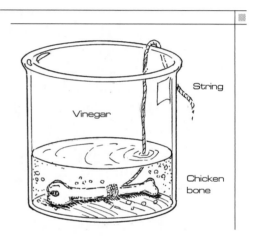

- Suspend the chicken bone in the vinegar for several days. Pull it out, then wash it well to remove any residual acid. Now show your students how the bone has become soft and rubbery. The acid in the vinegar has dissolved some of the minerals in the bone.

Gastric juices contain water, hydrochloric acid, pepsin, and rennin. Rennin is an enzyme that breaks down **casein,** or milk protein. The enzyme pepsin, together with hydrochloric acid, breaks down large protein molecules into smaller ones.

Materials: four test tubes, test tube holder, *diluted* hydrochloric acid (1:3), hard-boiled egg, water, *diluted* pepsin (1:3)

- Place four test tubes in a holder and label them #1, #2, #3, and #4.

- Place a small piece of egg white (a protein) in each tube and cover it with water. Test tube #1 will act as your control.

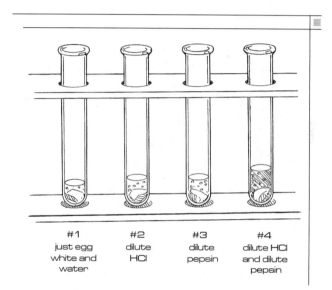

#1 just egg white and water #2 dilute HCl #3 dilute pepsin #4 dilute HCl and dilute pepsin

- Place 10 mL diluted hydrochloric acid in test tube #2.

- Place 10 mL pepsin diluted with water in test tube #3.

- Place 5 mL each diluted hydrochloric acid and diluted pepsin in test tube #4.

- Examine the four test tubes one day later. You will observe that the egg whites in #1 and #2 have not changed visibly. Some egg is left in #3, while most of the egg in #4 has been digested. This verifies that pepsin and hydrochloric acids are synergetic: They work better together than either one works alone.

Special Safety Consideration: Hydrochloric acid is corrosive. Use extreme caution when working with acids. Wear gloves, protective clothing, and safety goggles.

Heat is a key element in food digestion. In mammals and birds, metabolism rates are quite high, because mammals and birds are **endotherms** and require a consistent body temperature. Therefore, food must be digested as quickly as possible to support body functions. The heat generated by a mammalian or bird body assists rapid digestion.

Materials: three test tubes, *diluted* pepsin (1:3), *diluted* hydrochloric acid (1:3), incubator, refrigerator, hard-boiled egg

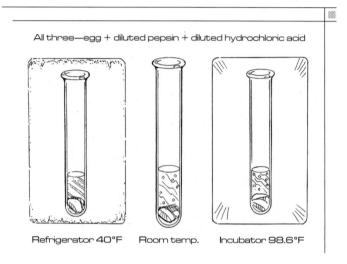

All three—egg + diluted pepsin + diluted hydrochloric acid

Refrigerator 40°F   Room temp.   Incubator 98.6°F

• Place nearly equal amounts of egg white in the three test tubes. Add to each tube equal amounts of diluted pepsin and diluted hydrochloric acid. Store one tube overnight in the refrigerator, one at room temperature, and one in the incubator at 37°C, or 98.6°F, human body temperature. You will be able to show students that the warmest tube shows the most digestion.

Special Safety Consideration: Hydrochloric acid is corrosive. Use extreme caution when working with acids. Wear gloves, protective clothing, and safety goggles.

Fats and oils are digested in the small intestine with the help of **lipase,** an enzyme. **Bile,** produced by the liver, is a greenish liquid stored in the **gall bladder.** Bile is not an enzyme but an emulsifier. Bile helps break down large fat drops into tiny fat droplets, dispersing them evenly through the small intestine (**emulsification**).

Materials: two test tubes, stoppers, water, salad oil, dropper, bile or liquid detergent (If bile is not available, use a couple of drops of liquid detergent in water.)

- Half-fill one test tube with bile. Half-fill the other tube with water. Add a couple of drops of oil to both; cover and shake. Notice how the oil floats in the water after you stop shaking, while the bile forms a cloudy mixture with the oil. Bile works in the same way during digestion in the small intestine.

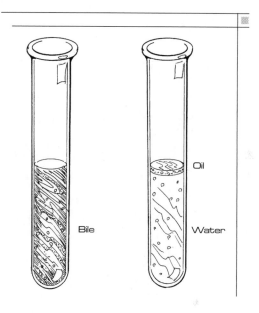

The small intestine is covered by millions of tiny fingerlike structures called **villi.** Villi contain capillaries and **lacteals,** which possess semipermeable membranes. Food products are absorbed through osmosis from the small intestine into the villi. The proteins are broken down into **amino acids,** and they and the sugars pass into the capillaries, while fats are passed through the lacteals. All food materials that pass through the capillaries and lacteals enter the bloodstream, where they can be used by the body. Fats are stored as reserve energy; proteins go to repair and build new cells; carbohydrates provide heat and muscle energy.

Materials: glass jar, coffee filter, water, salt or sugar

- Fill the glass jar with water. Place about a tablespoon (15 mL) of sugar or salt in the coffee filter. Fold the filter into a small bag and wet its upper end. Place the filter bag into the jar and fold the wet end of the bag over the lip of the jar.

You will observe a wavelike action coming out of the bag, almost like rain. This process is called osmosis. Substances move from a region of higher osmotic pressure to a region of lower pressure through a porous membrane. The sugar solution has more osmotic pressure than plain water, so it moves through the filter into the plain water. This demonstrates the action of villi.

Every time the heart beats, blood spurts into the arteries. If you feel an artery near a bone or near the surface of the skin, you can feel these spurts as **pulses.** The pulses and the heartbeat are the same. The normal heartbeat for a person at rest is 70 beats per minute. A mouse has a heartbeat of 1,000 beats per minute, while an elephant has one of 25 beats per minute. Generally, among mammals, the smaller the animal, the greater the number of heartbeats per minute.

Materials: **stethoscope,** watch with second hand, microphone, amplifier, speakers

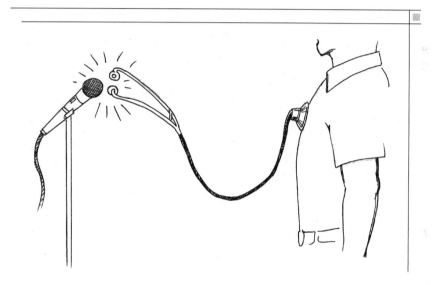

- Have a student place a stethoscope on his or her heart and have students listen to it. If you have a microphone with an amplifier and speakers, place the earpiece of the stethoscope near the microphone, and everyone will be able to hear the heart beating.

- Have students take their own pulses.

Nicotine, a strong and highly addictive stimulant found in cigarettes, increases a person's heartbeat rate.

1 jump
per second

2 jumps
per second

- Have students stand by their desks and move their chairs out of the way. Have them jump up once every second for about half a minute. Repeat the activity, but double the speed of jumps so that students jump up twice in each second. Students will quickly become nearly exhausted. Parallel the increase in jumping speed with the increase of heartbeats when nicotine is present. An increase of 30 heartbeats per minute means 43,200 extra heartbeats per day and 15,768,000 per year. These extra heartbeats place unneeded, damaging stress on the human heart muscle.

Materials: beaker, water, graduated cylinder, strainer or filter, marking pencil, eyedropper, nicotine solution, clock with second hand, guppy, overhead projector

In Preparation: Make the nicotine solution by soaking the tobacco of a cigarette in 15–20 milliliters of water for a couple of hours. Strain or filter the tobacco solution.

- Mark the front of the beaker with a vertical line. Fill the beaker with warm water and add the guppy. Let it swim around for a couple of minutes. Count and record how many times the fish swims around in one minute. Obtain the average of several tries. Place 18–20 drops of nicotine solution in the beaker water. Count how many times the fish swims around in one minute. Obtain the average of several tries. Rinse the fish in warm water. Promptly return the fish to its aquarium, so it will not be harmed. For clearer viewing, place the beaker on top of an overhead projector for bottom illumination.

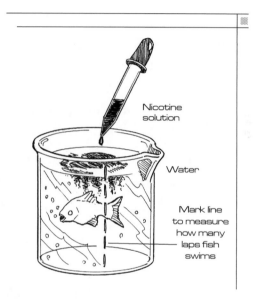

Nicotine solution

Water

Mark line to measure how many laps fish swims

Special Safety Consideration: Nicotine is toxic. It is used as a garden pesticide. In much higher concentrations, nicotine is fatal to humans. Dispose of nicotine solution immediately after use. Take care to make sure the guppy in this experiment is not harmed.

Indoor plants sometimes do poorly because of infestation by insects such as aphids. You will prepare a colloidal soap solution that will affect the **spiracles** of aphids. A **colloidal solution** consists of very fine particles of a substance suspended in another substance. Soap spreads the solution, for it is a good wetting agent. It breaks down the surface tension of the liquid and clogs the aphids' and other sucking insects' spiracles. In addition, nicotine is a very toxic chemical and kills eggs and **larvae** of insects.

Materials: plant infested with aphids, cigarette, water, tablespoon, small beaker, hot plate, measuring cup, liquid detergent, eyedropper, small sprayer

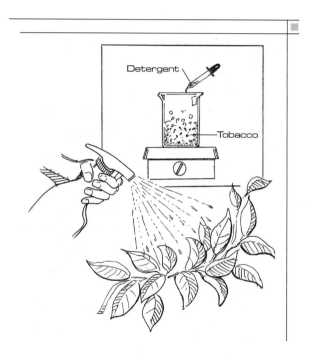

- Place 2–3 tablespoons (10–15 mL) of water in the beaker. Add the tobacco from a cigarette and boil for 8–10 minutes. Add enough water to make a cupful (225 mL) of solution. Add a few drops of liquid detergent. Mix and spray on the infested plant. Within a few days, students should observe a marked reduction in the level of infestation.

Special Safety Consideration: Nicotine is a toxic chemical. Wear gloves, goggles, and protective clothing.

When a crop is infested by pests, two choices are available to the grower: using pesticides or introducing a predator species to provide biological control. In this activity, a rose cutting infested with aphids will have biological control through the use of a lady bird beetle, better known as a ladybug. One of the ladybug's favorite foods is the aphid greenfly.

Materials: small bottle, soap, water, glass jar, rubber band, cheesecloth, aphid-infested rosebush cutting, small clippers, ladybug(s)

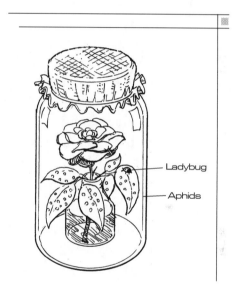

Ladybug

Aphids

- Wash both the bottle and the jar with soap and rinse well. Cut a rose with several aphid-infested leaves, about 4 inches in length. Place some water in the bottle, then insert the rose so that the leaves are outside the bottle. Place the bottle with the rose cutting inside the larger jar. Gently place a ladybug inside the jar. Cover the jar with cheesecloth and secure it with a rubber band. Observe for several days. You will notice a major drop in the aphid population. Release the ladybug outdoors at the end of the demonstration.

Special Note: Handle living animals with care. Even though aphids are what humans consider pest animals, they are entitled to the same respect as any other living creature.

Many people suffer from the lung disorder asthma. The word comes from Greek and means "panting." Asthma is usually caused by some irritant to the small air tubes in the lungs. To ease the symptoms of asthma, drugs can be used to relax the muscles of the air tubes. Emphysema, another lung disorder, occurs when the alveoli in the lungs break or become nonfunctional. It is often a side effect of smoking. For emphysema, an irreversible condition, oxygen is the only treatment. Breaths become gradually shallower (smaller) and use nearly all the energy of the afflicted individual.

Materials: drinking straw for each student

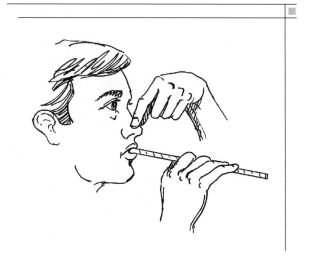

- Provide all students with a drinking straw and ask them to pinch their noses closed. Ask them to try to breathe only through the straw for three minutes. Warn them that if they feel faint or dizzy, they should stop at once. Monitor students carefully to watch for any signs of faintness. Explain to the students how this mimics emphysema and asthma.

Human body temperature is carefully regulated. The body sweats a combination of water and salt. The water evaporates, which cools the skin and maintains the body at 98.6°F, or 37°C. As liquids change to gas, they absorb heat energy (heat of vaporization). When air is saturated with water vapor (humidity), **perspiration** cannot **evaporate,** and individuals suffer from heat. Meteorologists use a comfort index, made up of the combination of temperature and moisture in the air, to arrive at a true relative temperature. This is the temperature as it is perceived by humans. Alcohol also needs heat to evaporate. While skin provides the heat, the sensation of alcohol on the skin is one of relative cooling.

Materials: dropper, isopropyl alcohol

- Have students line up. Place one drop of alcohol on the back of each student's hand. Have students take note of the cool feeling where you placed the alcohol drop.

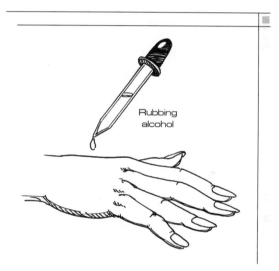

Rubbing alcohol

Many mammals (including human beings) have more than one type of hair on their bodies. For example, mammals with fur have "guard" hair that makes up the visible surface of the coat. This type of hair is thicker and coarser than the soft underhair that is closest to the animal's body.

Materials: hair from various mammals—gently cut hair from your dog, cat, rabbit, small rodent, or other pets or small barnyard animals (they will object less if you do this while you are holding them), microscope slides and coverslips, microscope, pen and paper

Procedure:

1. Make and label slides of each type of mammal hair you have been able to accumulate or your teacher has provided. Prepare the slide by dropping one drop of glycerine on a slide, placing the hair in the glycerine, and covering it with the coverslip.

2. Observe each slide under the microscope. Find the hair first under low power, then switch to high power.

3. Draw each and make brief descriptions of the differences.

Conclusion: How do hairs from different mammals compare? What is unique about each species, and what do they all have in common?

In the 1600s, Italian physician Francesco Redi observed the presence of worms on rotten meat. In time, he discovered that these were not really worms but the larvae of flies that had hatched from eggs. By placing gauze over the meat to prevent flies from reaching it, Redi showed that the meat did not produce worms or flies. He was able to demonstrate that flies come from flies. In other words, complex life does not arise spontaneously—there is no spontaneous generation.

Materials: two glass jars, rubber band, gauze, raw meat

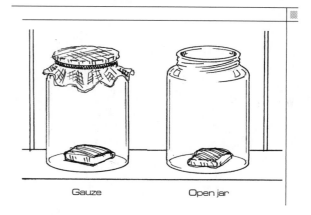

Gauze   Open jar

- Place a small amount of meat in each jar. Cover one jar with gauze, and leave the other one uncovered. Place both jars in a sunny window. After several days, examine the results. They should replicate Redi's findings.

Asexual reproduction means that only one organism is needed to produce an **offspring. Yeast** reproduces asexually by budding, or fission; that is, two cells of different size are produced. Yeast is a microscopic, one-celled fungus. In budding, the cell wall pushes out, beginning the bud. The cell nucleus moves toward the bud and divides, with one nucleus moving into the bud and the other remaining in the parent cell. The bud grows, and eventually a cell wall grows between the parent cell and the bud. Finally, the bud breaks away and develops into a mature cell.

Materials: flour, sugar, water, yeast, bread recipe, bowl, pan, microscope, slide, stain, eyedropper

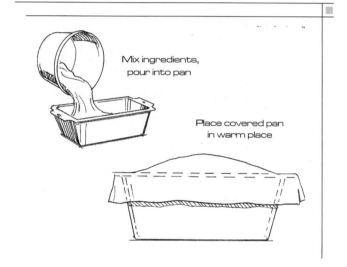

Mix ingredients, pour into pan

Place covered pan in warm place

- Following any standard bread recipe, mix all the ingredients in a bowl and then pour the dough into a pan. Cover the pan with a cloth, and set it in a warm place for most of the class period. Have students observe how the dough increases in size due to the action of the yeast.

- Begin the process of division by mixing a packet of dry yeast with a little sugar in a little warm water. Use the eyedropper to place a few drops of the yeast solution on a microscope slide. Stain to make the yeast easier to see, then place the slide under a microscope set at the highest power. Students will be able to see individual yeast dividing.

Bread mold under a microscope appears as many long stalks, each with a ball ending. Each ball is a spore case and contains thousands of cells called **spores.** Spores are the reproductive cells of molds. Each cell can grow into a new mold. The bread provides the nutrients for the growth of molds. When the ball that holds the spores (the spore case) breaks, **sporulation** (another form of asexual reproduction) takes place and the spores grow into new organisms.

Materials: bread, jar, two slides, microscope

- Place a piece of bread in the jar and let it stay uncovered until mold forms on the bread. This will take place in a couple of days. Prepare a slide and show students the mold and the spore cases.

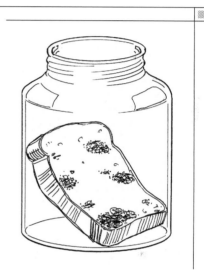

Special Safety Consideration: Many students have an allergic reaction to molds. Dispose of jars promptly, and ask if anyone is allergic to mold before allowing students to handle slides.

Sometimes, growing parts of plants develop into new plants. This form of asexual reproduction is called **vegetative propagation.** A potato is an underground stem known as a tuber. A potato grows many small eyes that, when planted, develop into new potato plants.

Materials: several potatoes, pot, soil, water, knife

- Let the potatoes develop eyes, then cut the eyes out with a large amount of potato around them. Plant them in the pot. Water lightly. Observe the growth of several new potato plants.

For asexual reproduction, only one parent is necessary. Seeds are the product of **sexual reproduction.** They are formed by the **fertilization** of an ovule by a pollen cell. After fertilization, the ovule becomes a seed. A new plant can grow from the seed. Germination is the process of growth for the embryo inside the seed.

Materials: large, flat beans, such as limas; water; iodine; eyedropper; jar or glass

- Take a seed, like a bean, and soak it overnight. Its coat will become soft and can be peeled off quite easily. The two halves will split naturally. They contain food, which is generally starch. (A quick iodine test will prove this.) Attached to one-half of the bean is the embryo.

- You can also use Demonstration 3, Phototropism #1: Plants Grow Toward Light. Have your students make daily observations to see how the seeds sprout.

Labels: Embryonic leaves, Embryonic stem, Embryonic root, Seed coat, Cotyledons

Fruit flies are a good example of rapid insect multiplication. If left unchecked, insects would overrun the world, but then so would any other population left unchecked.

Materials: two pint or quart jars with covers, grapes and/or ripe banana, cotton covering

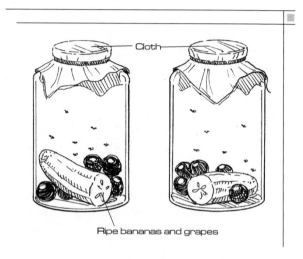

Cloth

Ripe bananas and grapes

- Place a piece of ripe banana and/or several grapes in both jars and let them sit in the open until fruit flies appear; flies may have either entered the jar or hatched from eggs on the fruit. Cover the jars with cloth and place in a warm room, not in sunlight. Keep track of the flies in each bottle daily for 8–10 days. Students should observe and record the increase of fruit flies, then calculate and project the increase over months and years. The numbers will be astounding.

Note: After observing insects for a time, release them to the outdoors.

The life cycle of a fruit fly is so short that a few weeks can produce multiple generations. Larvae emerge from eggs in one or two days, then pupate for about five days to change into adults and mate. Female fruit flies are larger than males and have a bigger abdomen. Males have a black-tipped abdomen.

Materials: one or two glass jars, ripe banana or other fruit, crumpled paper, magnifying glass, test tubes, foil

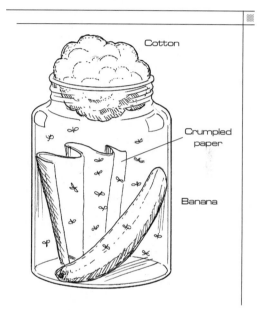

Cotton

Crumpled paper

Banana

- Place the ripe fruit in the jar and let it sit in the open until fruit flies appear; flies may have either entered the jar or hatched from eggs on the fruit. Place a piece of crumpled paper inside the jar, then cover the jar with absorbent cotton. You will observe larvae, **pupae,** and adult fruit flies in less than two weeks. The jar will contain the fruit flies that were caught and the new generation of flies. You can continue by preparing another habitat jar and raising a third generation. To observe fruit flies closely, place them in test tubes. Wrap the test tubes in foil and open small areas for observation. The fruit flies will come to the openings, attracted by the light. Use the magnifying glass to observe details.

Note: Release all fruit flies to the outdoors when observations are complete.

The general appearance of living organisms provides clues to many inherited characteristics called **traits.** While families share many traits, specific characteristics are uniquely individual. Examples of **inherited traits** include height (short, medium, tall), hair color (brown, black, blond, red), eye color (brown, blue, hazel), and frame size (small, medium, large).

Materials: family pictures
   (Ask each student to bring in one.)

- Have students write out as many traits as possible that are common in their family.

Traits are inherited through **genes.** Genes represent an organism's total inherited potential, if the environment allows for the genes' full expression. When the environment is restricted, the genes do not change, but the development of traits is stunted.

Materials: two small potted plants (raise these plants from seeds or cuttings from the same parent plant—the potato plants in Demonstration 58 will work fine), water, fertilizer

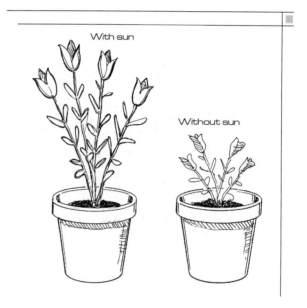

With sun

Without sun

- Take two small plants and let one develop to its full potential with abundant sunshine, water, and fertilizer. Give the second plant the same amount of water and fertilizer, but keep it in the shade. You will notice a major discrepancy in traits. The difference is due to the environment, not to genetics, because both plants come from the same parent.

Fish, like other living things, are adapted to prefer certain colors, moisture levels, temperatures, and other variables. These adaptations are reflected by the **habitats** in which they live.

Materials: fish tank, water, several varieties of fish, colored cards or papers, tape

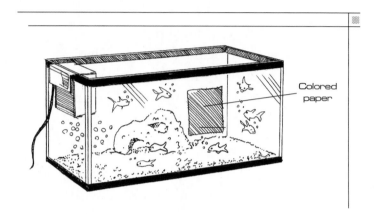

Colored paper

- Select one species of fish and place some in the tank, after creating a suitable habitat. Tape a colored card or paper to the outside of the glass tank and have students count the number of fish that congregate near it over a set time. Try other colors. Are certain colors attractive to this particular species?

- Repeat this demonstration with different species of fish.

Note: Once you have purchased fish for this experiment, you must then take responsibility for their welfare afterward. Make sure that the fish you select can thrive in a community tank at a particular temperature so that you do not have to set up three or four aquariums in the classroom!

Moths, as well as many other nocturnal insects, are attracted by your porch light at night. You are aware that some animals are nocturnal and others are diurnal. In this experiment, we will capture and identify insects in your area that are not only nocturnal, but are also attracted to bright lights.

Materials: flashlight, paper bag, tape, magnifying glass, field guide for insects, access to a freezer

Procedure:

1. Tape a paper bag over a flashlight. Cut a small hole in the end of the bag opposite the flashlight (about 5 cm in diameter). Turn on the flashlight, and leave it in a place where insects congregate at night (your front yard, for instance).

2. Leave it on for several hours after dark.

3. When you have trapped several insects, cover the opening with tape and place the trap in the freezer for an hour. This will temporarily immobilize the insects without permanently harming them.

4. Empty the trap, and examine the insects with a magnifying glass. Count the number of each kind and identify them.

Conclusion: How is nocturnal life an adaptation for some insects? Why are some insects attracted to bright lights?

Note: Release insects after observations.

All living plants and animals are adapted to prefer certain colors, moisture levels, temperatures, and other life-affecting variables. Their habitat causes some of these adaptations; in other cases, animals may migrate in search of a more hospitable habitat elsewhere.

In the following demonstration, we will work with simple earthworms. Earthworms are **annelids** and spend the majority of their lives underground. Thus, they are adapted to a dark, cool, moist environment.

Materials: worms, paper of different colors, water

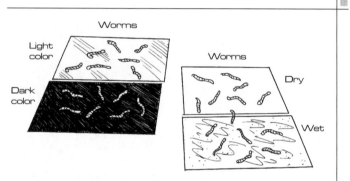

- Place the worms on two cards of different colors, one light and one dark. Notice which card most worms migrate toward.

- Place the worms on two cards of the same color, one moist, one dry. Observe the mass migration to one of the cards.

- This demonstration can be repeated to test for many other variables, such as mild acidity, mild baseness, odor, temperature, etc.

Note: It is important to handle live animals carefully to avoid harming them inadvertently. Release the worms outdoors after demonstrations, if possible, or keep them in a safe environment indoors, such as a warm box or wormery, which you will be assembling in the next demonstration.

Earthworms, members of the annelid phylum, move through earth. They do this by making the front of their segmented bodies longer and contracting the back parts. Since earthworms live underground, we rarely see them moving. In this activity, you will build a wormery so students can observe how earthworms crawl around.

Materials: damp soil; large, empty glass jar (no lid needed); clean sand; leaves; garden spade; three or four earthworms; rubber band; cheesecloth; black construction paper; sealing tape

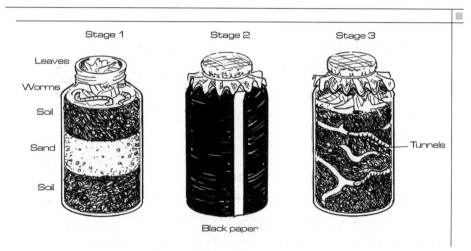

* Place a layer of damp soil on the bottom of the jar. Add to this a layer of clean sand, then add more damp soil. Using the spade, dig in moist, warm spots in a garden to unearth several earthworms. Collect a few green leaves. Place the worms and the leaves in the jar on top of the soil. Cover the jar with cheesecloth, secured with the rubber band. Wrap the entire jar in black construction paper, making certain to overlap the edges. Seal it with tape. Set the jar in a safe place. After a few days, remove the black paper and observe the worm burrows. You will see tunnels through the soil and the sand.

Note: Handle living animals carefully to avoid harming them inadvertently.

Earthworms like a moist, dark environment. They will tunnel into the soil only after everything is dark. In the process of tunneling, they pass the soil through their bodies, remove the nutrients from the organic materials in the soil, and deposit a substance called castings. Earthworm tunnels aerate the soil, and the castings enrich it.

Materials: tall, thin aluminum can (open at one end); large clear glass or plastic jar; earthworms; rich soil; sand; water; black construction paper; tape

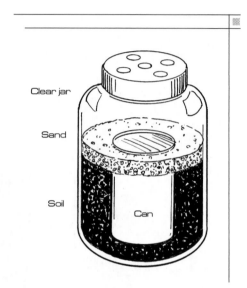

Clear jar

Sand

Soil

Can

• Have your students look at an earthworm and try to decide which end is the head and which end is the tail. Place the can inside the jar, closed end up. (The can is introduced so that earthworms are forced to tunnel near the surfaces of the clear jar.) Fill the jar with the soil up to the top of the can. Add a thin layer of sand on top of the soil. Moisten the soil but do not overwater. Place the earthworms in the jar. Wrap the entire jar in black paper, making sure the ends overlap. Make some airholes in the cover of the jar, and seal the jar. Leave the jar for several days. When you remove the black paper, you will observe worm castings on top of the layer of sand in the jar.

Note: Handle living animals carefully to avoid harming them inadvertently.

Nematodes are widely distributed worms, sometimes called roundworms (to distinguish them from flatworms). Some are parasitic to humans, animals, or plants, while others are harmless. Parasitic species include trichinae, hookworms, and pinworms.

However, many species are free-living and feed on bacteria in moist soil. They recycle soil nutrients and destroy some agricultural pests. One square meter of garden soil will yield millions of nematodes. In this lab, we will be examining garden soil with a magnifying glass to find nematodes and compare their movements to those of annelids, or earthworms.

Materials: magnifying glass, 10 cc garden soil, toothpicks, earthworms, colored paper

Procedure:

1. Place a small amount of soil on a sheet of colored paper on your desk.

2. While observing with the magnifying glass, pull apart any clumps in the soil with toothpicks.

3. Nematodes are small, white threadlike animals, wriggling rapidly.

4. Place the earthworm on your paper. How is the earthworm's movement different from the nematode's?

Conclusion: How many nematodes did you find? How did they move? How are they well-adapted to their environment? What were the major differences between the nematodes and the earthworms? How is the earthworm adapted to its environment?

Note: Handle living animals carefully to avoid harming them inadvertently.

Many species of insects go through a complete **metamorphosis,** where their larvae bear a resemblance to true worms. This is true of mealworms. A mealworm is the larva of one of many grain-eating beetles. There is a pupa stage as well, which lasts a week or so and looks something like a small, shiny black pill. Metamorphosis is clearly demonstrated by mealworms changing into beetles.

Materials: glass jar, clear plastic, grain cereal (cornmeal works well), mealworms (available at many pet stores as food for other small pets, such as geckos)

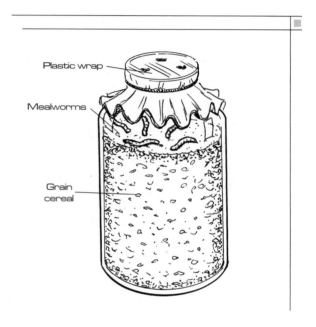

- Put grain cereal in a jar and place the mealworms on it. Cover the jar with plastic and stand it in plain view. Do not forget to poke several airholes in the plastic. Depending on the room temperature, the mealworms will change into beetles in one or two weeks. This is a good example of metamorphosis with a commonly available animal.

Note: When the animals have completed their metamorphosis, release them outdoors.

In the average drop of pond water, thousands of separate organisms can be seen. Paramecia, rotifers, algaes, ordinary bacteria, and long chains of cyanobacteria can easily be observed by students. This demonstration is more effective if you have a microscope projector; however, you can also simply set up several microscopes with prepared slides.

Materials: 1 liter of pond water, eyedropper, slides, coverslips, one microscope with projecting capability or several ordinary microscopes, a prepared guide to pond water life (see below)

- Prepare slides by carefully placing one drop of water on each slide and covering with a slip. Place under a microscope at low to medium power. Observe, with students, what sorts of life forms can be seen under the microscope. Identify as many as possible using the simple guide below.

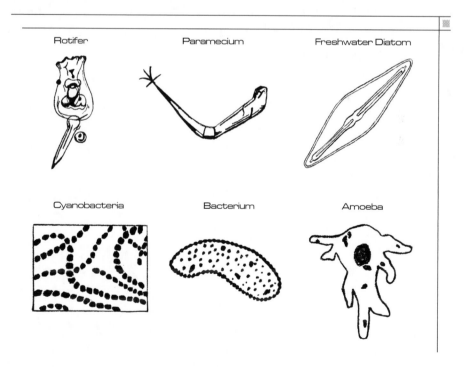

Rotifer   Paramecium   Freshwater Diatom

Cyanobacteria   Bacterium   Amoeba

Special Safety Consideration: Some bacteria and other pond life forms can cause illness if ingested. Wear gloves and goggles, and wash hands thoroughly after handling pond organisms.

During their lifetimes, plants use up only traces of minerals from the soil. Their extensive growth is made possible by their production of food from sunlight, water, and carbon dioxide.

Materials: small flowerpot, potting soil, seeds, water, balance

- Determine the mass of soil needed to fill the pot. Find the mass of the seeds. Plant the seeds in the potting soil and water them. Wait until the plants are several inches tall. Remove them, making sure that you brush off into the pot all the soil from their roots. Now measure the mass of the soil and the plants again. Compare their masses with the original figures.

Hydroponics is the science of growing plants primarily in a nutrient solution in water, without soil. This method of plant growth was discovered in the 1800s when scientists were trying to study the nutritional needs and root systems of plants. There are two types of hydroponics: water culture and aggregate culture. In a water culture, the plants are suspended over a bed of nutrient solution. As the plants grow, the roots reach into the nutrient solution. In an aggregate culture, the plants are held in coarse materials such as sand and gravel, and the nutrient solution is constantly circulated in the anchoring materials. You might want to try out different combinations of nutrients as your project grows. Tomatoes do particularly well in hydroponics.

Materials: aquarium, wire mesh, wire mesh cutter, water, $2\frac{3}{4}$ gallons (10.4 L) distilled water, small sack sphagnum moss (peat moss), 2 teaspoons (10 mL) calcium nitrate, $\frac{1}{4}$ teaspoon (1.25 mL) ammonium nitrate, $\frac{1}{4}$ teaspoon (1.25 mL) ammonium sulfate, $1\frac{1}{4}$ teaspoons (6.25 mL) Epsom salts, $\frac{1}{8}$ teaspoon (.625 mL) potassium acid phosphate, $\frac{1}{8}$ teaspoon (.625 mL) boric acid, $\frac{1}{8}$ teaspoon (.625 mL) manganese sulfate, $\frac{1}{8}$ teaspoon (.625 mL) zinc sulfate, $\frac{1}{8}$ teaspoon (.625 mL) ferrous sulfate, bean and corn seeds, measuring cup, set of measuring spoons or graduated cylinders, stirrer

- Follow these instructions carefully.

1. Cut the wire mesh so that it is as wide as the aquarium and 8 inches longer. Four inches from each end, bend the wire mesh at right angles to form a platform the same length as the aquarium, with 4-inch uprights. Place the mesh platform inside the tank. The uprights should keep it about 4 inches above the base of the tank. (See diagram.)

2. Prepare three solutions: #1, #2, and #3. Only #1 will go into the aquarium, at the very end.

(continued)

Make Solution #1 by dissolving in a cup (235 mL) of water:

$\frac{1}{4}$ teaspoon (1.25 mL) ammonium sulfate

$\frac{1}{2}$ teaspoon (2.5 mL) potassium acid phosphate

$1\frac{1}{4}$ teaspoon (6.25 mL) Epsom salts

2 teaspoons (10 mL) calcium nitrate

When all the above are dissolved, add the solution to $2\frac{1}{2}$ gallons (9.46 L) of distilled water and label it Solution #1.

Make Solution #2 by dissolving in a cup (235 mL) of water:

$\frac{1}{8}$ teaspoon (.625 mL) zinc sulfate

$\frac{1}{8}$ teaspoon (.625 mL) manganese sulfate

$\frac{1}{8}$ teaspoon (.625 mL) boric acid

When all are dissolved, label it Solution #2.

Make Solution #3 by dissolving in a cup (235 mL) of water:

$\frac{1}{8}$ teaspoon (.625 mL) ferrous sulfate

When it is dissolved, label it Solution #3.

3. Add 1 teaspoon (5 mL) of Solution #2 to Solution #1.

4. Add 3 tablespoons (45 mL) of Solution #3 to Solution #1.

5. Pour the final Solution #1 (mixture) into the aquarium up to the level of the wire mesh platform.

6. Sprinkle a layer of peat moss (sphagnum moss) over the wire mesh platform. Place bean and corn seeds on top of the peat moss.

(continued)

Over the next few days, the seeds will germinate, and their roots will reach into the nutrient solution. Maintain the level of nutrients by adding Solution #1 to keep it to the level of the wire platform. If needed, make more solution.

Special Safety Consideration:
Ammonium nitrate is a rapid oxidizer and poses a minor explosion hazard if exposed to flames or sparks. Keep ammonium nitrate well away from students.

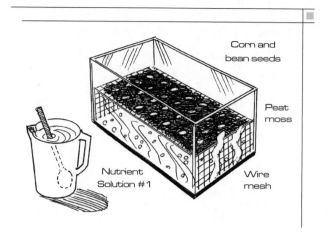

Corn and bean seeds

Peat moss

Nutrient Solution #1

Wire mesh

A mold is a simple organism belonging to Kingdom Fungi. Since molds cannot make their own food, they exist as parasites on plants and animals. Molds are useful in many ways. Many cheeses depend on molds for their ripening. Blue cheese is one example. Molds are present in many fertilizers. Antibiotics, such as penicillin, are derived from molds. In this demonstration, you will grow a variety of molds.

Materials: three clean glass jars with covers, plate, slice of bread, bruised pear, bruised apple, water

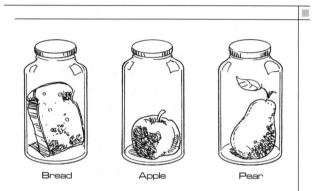

Bread   Apple   Pear

- Place the slice of bread on a plate and leave it in the air for about an hour. Mold spores will fall on it from the air. Then place the slice of bread in a jar, sprinkle it lightly with water, and close the jar. Place the bruised apple in a jar and close it. Do the same with the pear. Place the three jars in a warm, dark place. After one week, bring them out and examine them. The molds working on the fruits will have rotted them. The bread will be covered with a cottonlike growth of white stalks with black balls at the ends. These are spore cases. Following the examination, discard the unopened jars.

Special Safety Consideration: Many students are allergic to mold spores if released to the air. Do not open the jars during examination.

Erosion is the process of breaking down and carrying away the materials of earth, such as rocks and soil. Most erosion takes place slowly. Thousands of years of glacial erosion have created many North American mountain ranges. The Grand Canyon is a spectacular result of erosion. The Colorado River carved out its contours over millions of years. Erosion starts with the process of weathering. During weathering, earth materials are broken down into smaller pieces. The movement of water and air and the heat of the sun all contribute to weathering. Once the materials are loosened, water and wind can carry them to new locations. Heavy winds can blow soil and rocky substances over great distances, and heavy rains can wash soil particles downhill. Intensive farming can be an additional cause of soil erosion. When the soil is cleared of plants and trees, all its shields from wind and rain are removed. Then the topsoil is at risk of being washed or blown away. To limit the erosion of topsoil, farmers plant cover crops, such as alfalfa or grass, or use tillage of the soil. Tilling allows old crops to remain on the surface of the soil.

In this demonstration you will observe some common approaches to reducing soil erosion.

Materials: scrap pieces of wood (2 × 4), plastic or rubber tubing, florist's clay, scissors, topsoil, sand, three glass or plastic bowls, three aluminum foil cake pans, cereal grain seeds, sprinkling can, water

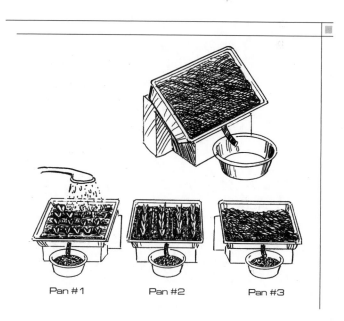

Pan #1   Pan #2   Pan #3

• Use the scissors to cut a small hole near the top of each of the three aluminum foil cake pans, in the middle of the longer side. Insert in each hole a 10-inch piece of tubing, and seal it with florist's clay. The holes can be larger than the tubing. Fill each pan with a layer of clay, a layer of topsoil, a layer of sand, and

(continued)

a final layer of topsoil. With the hose pointing down, place the three pans at an angle of about 30° by placing them over a block, with another rear support block. Place the hoses over the bowls to collect the runoff from the pans. Next, plant your crops of cereal grains, such as wheat, oats, or barley. In the first pan, plant the seeds in horizontal rows. Plant the seeds in the second pan in vertical rows. Do not plant anything in the third pan. Wait until your young plants have germinated and are well established, watering gently. Now simulate rain by watering the three pans equally, using the same amount of water for each. Observe the topsoil in the pans for the next few days. The horizontal rows duplicate the contour farming that farmers use around a hill. The vertical method, up and down a hill, prevents some soil erosion. Planting nothing at all leaves the soil unprotected. Contour farming, as in pan #1, appears to be the best way to prevent topsoil erosion.

Molecules are the smallest parts of a substance that have the same properties as the substance itself. Molecules of gases, liquids, and solids are continuously in motion. **Diffusion** is the mixing of molecules of one substance with those of another one. Molecular motion causes diffusion.

Materials: four glasses, water, sugar cube, hard candy, rock salt, dark food coloring, bottle of perfume, or coffeepot and coffee

Sugar cube   Hard candy   Rock salt

Food coloring

- Make a pot of coffee or open a bottle of perfume, and observe how rapidly the aroma of coffee or perfume diffuses throughout the room.

- Half-fill three glasses with water. Place the sugar cube in one, the candy in the next, and the piece of rock salt in the last one. Do not stir. Diffusion causes the dissolving (mixing) of these materials into water.

- Fill one glass nearly full. Let it stand for a few minutes. Carefully place in it one drop of a dark food coloring. Observe it for eight to ten minutes. Diffusion takes place due to molecular motion.

Knowing that acids can break apart rocks and the surface of the earth, it is interesting to realize that the roots of plants are acidic. We can test for acids and bases easily by using litmus-paper indicator strips.

Materials: dish with glass cover, moist cotton, distilled water, several seeds, blue litmus paper

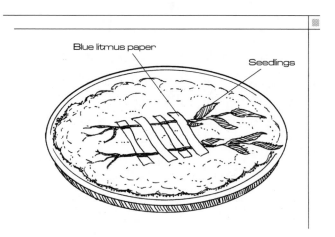

Blue litmus paper

Seedlings

- Place several seeds on the moist cotton in the dish, cover, and let the seeds germinate into small plants. When the seedlings have grown a good root system, you will be ready to run the test for acid. Moisten several strips of blue litmus paper with distilled water and place them over and below the rootlets. Observe the change in the color of the litmus paper over a couple of days. (Make certain that the cotton is not too wet, or the test will not work.)

Appendix

1. Assessing Laboratory Reports

This book contains 10 student laboratory assignments, for which students should be expected to produce a written report. Go over what you want to see in a lab report with your students before they start. Information should include:

- **Purpose:** Why is this lab being performed? What is the objective of the lab?

- **Hypothesis:** Given the initial level of knowledge, what do students expect to find out at the end?

- **Materials list:** Students should be told that one of the main reasons for writing lab reports at all is so the labs can be replicated by others. A well-organized materials list makes it easier for you to understand what they did, and for those who might try to replicate their results to do so.

- **Procedure:** Likewise, students should include the procedure they and their lab partners actually followed.

- **Data:** What actually took place in the lab? In life sciences, observations of changes in behavior, as well as observations of organism response to stimuli, are important, and if students observed any of these things, they should be recorded.

- **Conclusion:** What were the results? Did their hypotheses match the data? If something went wrong, what does the student think happened?

(continued)

1. Assessing Laboratory Reports *(continued)*

In order to give you a quick guide to assessing lab reports, we have constructed the following rubric:

	1	2	3	4
Understanding of Concept	Poor	Adequate	Good	Outstanding
Methodology	Poor	Adequate	Good	Outstanding
Organization of Experiment	Poor	Adequate	Good	Outstanding
Organization of Report	Poor	Adequate	Good	Outstanding

Laboratory reports are an important stepping stone for young scientists, but they can become burdensome to correct. We hope this rubric assists the typical busy teacher in providing a quality lab experience without sacrificing the deeper knowledge of the scientific method that writing lab reports reinforces among young scientists.

Easy Science Demos & Labs:
Life Science

2. Density of Liquids

approx. gm/cm^3
at 20°C

Acetone	0.79
Alcohol (ethyl)	0.79
Alcohol (methyl)	0.81
Benzene	0.90
Carbon disulfide	1.29
Carbon tetrachloride	1.56
Chloroform	1.50
Ether	0.74
Gasoline	0.68
Glycerin	1.26
Kerosene	0.82
Linseed oil (boiled)	0.94
Mercury	13.6
Milk	1.03
Naphtha (petroleum)	0.67
Olive oil	0.92
Sulfuric acid	1.82
Turpentine	0.87
Water 0°C	0.99
Water 4°C	1.00
Water–sea	1.03

3. Altitude, Barometer, and Boiling Point

altitude (approx. ft)	barometer reading (cm of mercury)	boiling point (°C)
15,430	43.1	84.9
10,320	52.0	89.8
6,190	60.5	93.8
5,510	62.0	94.4
5,060	63.1	94.9
4,500	64.4	95.4
3,950	65.7	96.0
3,500	66.8	96.4
3,060	67.9	96.9
2,400	69.6	97.6
2,060	70.4	97.9
1,520	71.8	98.5
970	73.3	99.0
530	74.5	99.5
0	76.0	100.0
−550	77.5	100.5

4. Specific Gravity

gram/cm^3 at 20°C

Agate	2.5–2.6	Granite*	2.7	Polystyrene	1.06
Aluminum	2.7	Graphite	2.2	Quartz	2.6
Brass*	8.5	Human body–normal	1.07	Rock salt	2.1–2.2
Butter	0.86	Human body–lungs full	1.00	Rubber (gum)	0.92
Cellural cellulose acetate	0.75	Ice	0.92	Silver	10.5
Celluloid	1.4	Iron (cast)*	7.9	Steel	7.8
Cement*	2.8	Lead	11.3	Sulfur (roll)	2.0
Coal (anthracite)*	1.5	Limestone	2.7	Tin	7.3
Coal (bituminous)*	1.3	Magnesium	1.74	Tungsten	18.8
Copper	8.9	Marble*	2.7	Wood: Rock Elm	0.76
Cork	0.22–0.26	Nickel	8.8	Balsa	0.16
Diamond	3.1–3.5	Opal	2.1–2.3	Red Oak	0.67
German silver	8.4	Osmium	22.5	Southern Pine	0.56
Glass (common)	2.5	Paraffin	0.9	White Pine	0.4
Gold	19.3	Platinum	21.4	Zinc	7.1

*Non-homogeneous material. Specific gravity may vary. Table gives average value.

© 1996, 2003
J. Weston Walch, Publisher

Easy Science Demos & Labs:
Life Science

5. Temperature Conversion (Celsius to Fahrenheit)

C°	F°	C°	F°	C°	F°	C°	F°	C°	F°	C°	F°
250	482.00	200	392.00	150	302.00	100	212.00	50	122.00	0	32.00
249	480.20	199	390.20	149	300.20	99	210.20	49	120.20	−1	30.20
248	478.40	198	388.40	148	298.40	98	208.40	48	118.40	−2	28.40
247	476.60	197	386.60	147	296.60	97	206.60	47	116.60	−3	26.60
246	474.80	196	384.80	146	294.80	96	204.80	46	114.80	−4	24.80
245	473.00	195	383.00	145	293.00	95	203.00	45	113.00	−5	23.00
244	471.20	194	381.20	144	291.20	94	201.20	44	111.20	−6	21.20
243	469.40	193	379.40	143	289.40	93	199.40	43	109.40	−7	19.40
242	467.60	192	377.60	142	287.60	92	197.60	42	107.60	−8	17.60
241	465.80	191	375.80	141	285.80	91	195.80	41	105.80	−9	15.80
240	464.00	190	374.00	140	284.00	90	194.00	40	104.00	−10	14.00
239	462.20	189	372.20	139	282.20	89	192.20	39	102.20	−11	12.20
238	460.40	188	370.40	138	280.40	88	190.40	38	100.40	−12	10.40
237	458.60	187	368.60	137	278.60	87	188.60	37	98.60	−13	8.60
236	456.80	186	366.80	136	276.80	86	186.80	36	96.80	−14	6.80
235	455.00	185	365.00	135	275.00	85	185.00	35	95.00	−15	5.00
234	453.20	184	363.20	134	273.20	84	183.20	34	93.20	−16	3.20
233	451.40	183	361.40	133	271.40	83	181.40	33	91.40	−17	1.40
232	449.60	182	359.60	132	269.60	82	179.60	32	89.60	−18	−0.40
231	447.80	181	357.80	131	267.80	81	177.80	31	87.80	−19	−2.20
230	446.00	180	356.00	130	266.00	80	176.00	30	86.00	−20	−4.00
229	444.20	179	354.20	129	264.20	79	174.20	29	84.20	−21	−5.80
228	442.40	178	352.40	128	262.40	78	172.40	28	82.40	−22	−7.60
227	440.60	177	350.60	127	260.60	77	170.60	27	80.60	−23	−9.40
226	438.80	176	348.80	126	258.80	76	168.80	26	78.80	−24	−11.20
225	437.00	175	347.00	125	257.00	75	167.00	25	77.00	−25	−13.00
224	435.20	174	345.20	124	255.20	74	165.20	24	75.20	−26	−14.80
223	433.40	173	343.40	123	253.40	73	163.40	23	73.40	−27	−16.60
222	431.60	172	341.60	122	251.60	72	161.60	22	71.60	−28	−18.40
221	429.80	171	339.80	121	249.80	71	159.80	21	69.80	−29	−20.20
220	428.00	170	338.00	120	248.00	70	158.00	20	68.00	−30	−22.00
219	426.20	169	336.20	119	246.20	69	156.20	19	66.20	−31	−23.80
218	424.40	168	334.40	118	244.40	68	154.40	18	64.40	−32	−25.60
217	422.60	167	332.60	117	242.60	67	152.60	17	62.60	−33	−27.40
216	420.80	166	330.80	116	240.80	66	150.80	16	60.80	−34	−29.20
215	419.00	165	329.00	115	239.00	65	149.00	15	59.00	−35	−31.00
214	417.20	164	327.20	114	237.20	64	147.20	14	57.20	−36	−32.80
213	415.40	163	325.40	113	235.40	63	145.40	13	55.40	−37	−34.60
212	413.60	162	323.60	112	233.60	62	143.60	12	53.60	−38	−36.40
211	411.80	161	321.80	111	231.80	61	141.80	11	51.80	−39	−38.20
210	410.00	160	320.00	110	230.00	60	140.00	10	50.00	−40	−40.00
209	408.20	159	318.20	109	228.20	59	138.20	9	48.20	−41	−41.80
208	406.40	158	316.40	108	226.40	58	136.40	8	46.40	−42	−43.60
207	404.60	157	314.60	107	224.60	57	134.60	7	44.60	−43	−45.40
206	402.80	156	312.80	106	222.80	56	132.80	6	42.80	−44	−47.20
205	401.00	155	311.00	105	221.00	55	131.00	5	41.00	−45	−49.00
204	399.20	154	309.20	104	219.20	54	129.20	4	39.20	−46	−50.80
203	397.40	153	307.40	103	217.40	53	127.40	3	37.40	−47	−52.60
202	395.60	152	305.60	102	215.60	52	125.60	2	35.60	−48	−54.40
201	393.80	151	303.80	101	213.80	51	123.80	1	33.80	−49	−56.20

6. Temperature Conversion (Fahrenheit to Celsius)

F°	C°	F°	C°	F°	C°	F°	C°	F°	C°	F°	C°
250	121.11	200	93.33	150	65.56	100	37.78	50	10.00	0	−17.78
249	120.56	199	92.78	149	65.00	99	37.22	49	9.44	−1	−18.33
248	120.00	198	92.22	148	64.44	98	36.67	48	8.89	−2	−18.89
247	119.44	197	91.67	147	63.89	97	36.11	47	8.33	−3	−19.44
246	118.89	196	91.11	146	63.33	96	35.56	46	7.78	−4	−20.00
245	118.33	195	90.56	145	62.78	95	35.00	45	7.22	−5	−20.55
244	117.78	194	90.00	144	62.22	94	34.44	44	6.67	−6	−21.11
243	117.22	193	89.44	143	61.67	93	33.89	43	6.11	−7	−21.67
242	116.67	192	88.89	142	61.11	92	33.33	42	5.56	−8	−22.22
241	116.11	191	88.33	141	60.56	91	32.78	41	5.00	−9	−22.78
240	115.56	190	87.78	140	60.00	90	32.22	40	4.44	−10	−23.33
239	115.00	189	87.22	139	59.44	89	31.67	39	3.89	−11	−23.89
238	114.44	188	86.67	138	58.89	88	31.11	38	3.33	−12	−24.44
237	113.89	187	86.11	137	58.33	87	30.56	37	2.78	−13	−25.00
236	113.33	186	85.56	136	57.78	86	30.00	36	2.22	−14	−25.56
235	112.78	185	85.00	135	57.22	85	29.44	35	1.67	−15	−26.11
234	112.22	184	84.44	134	56.67	84	28.89	34	1.11	−16	−26.67
233	111.67	183	83.89	133	56.11	83	28.33	33	0.56	−17	−27.22
232	111.11	182	83.33	132	55.56	82	27.78	32	0.00	−18	−27.78
231	110.56	181	82.78	131	55.00	81	27.22	31	−0.56	−19	−28.33
230	100.00	180	82.22	130	54.44	80	26.67	30	−1.11	−20	−28.89
229	109.44	179	81.67	129	53.89	79	26.11	29	−1.67	−21	−29.44
228	108.89	178	81.11	128	53.33	78	25.56	28	−2.22	−22	−30.00
227	108.33	177	80.56	127	52.78	77	25.00	27	−2.78	−23	−30.56
226	107.78	176	80.00	126	52.22	76	24.44	26	−3.33	−24	−31.11
225	107.22	175	79.44	125	51.67	75	23.89	25	−3.89	−25	−31.67
224	106.67	174	78.89	124	51.11	74	23.33	24	−4.44	−26	−32.22
223	106.11	173	78.33	123	50.56	73	22.78	23	−5.00	−27	−32.78
222	105.56	172	77.78	122	50.00	72	22.22	22	−5.56	−28	−33.33
221	105.00	171	77.22	121	49.44	71	21.67	21	−6.11	−29	−33.89
220	104.44	170	76.67	120	48.89	70	21.11	20	−6.67	−30	−34.44
219	103.89	169	76.11	119	48.33	69	20.56	19	−7.22	−31	−35.00
218	103.33	168	75.56	118	47.78	68	20.00	18	−7.78	−32	−35.56
217	102.78	167	75.00	117	47.22	67	19.44	17	−8.33	−33	−36.11
216	102.22	166	74.44	116	46.67	66	18.89	16	−8.89	−34	−36.67
215	101.67	165	73.89	115	46.11	65	18.33	15	−9.44	−35	−37.22
214	101.11	164	73.33	114	45.56	64	17.78	14	−10.00	−36	−37.78
213	100.56	163	72.78	113	45.00	63	17.22	13	−10.56	−37	−38.33
212	100.00	162	72.22	112	44.44	62	16.67	12	−11.11	−38	−38.89
211	99.44	161	71.67	111	43.89	61	16.11	11	−11.67	−39	−39.44
210	98.89	160	71.11	110	43.33	60	15.56	10	−12.22	−40	−40.00
209	98.33	159	70.56	109	42.78	59	15.00	9	−12.78	−41	−40.56
208	97.78	158	70.00	108	42.22	58	14.44	8	−13.33	−42	−41.11
207	97.22	157	69.44	107	41.67	57	13.89	7	−13.89	−43	−41.67
206	96.67	156	68.89	106	41.11	56	13.33	6	−14.44	−44	−42.22
205	96.11	155	68.33	105	40.56	55	12.78	5	−15.00	−45	−42.78
204	95.56	154	67.78	104	40.00	54	12.22	4	−15.56	−46	−43.33
203	95.00	153	67.22	103	39.44	53	11.67	3	−16.11	−47	−43.89
202	94.44	152	66.67	102	38.89	52	11.11	2	−16.67	−48	−44.44
201	93.89	151	66.11	101	38.33	51	10.56	1	−17.22	−49	−45.00

Easy Science Demos & Labs:
Life Science

Glossary

A

adhesion: when molecules of one kind stick to molecules of other kinds

aerobic organism: an organism that obtains its oxygen from the air or water

amino acids: building blocks of proteins; during digestion, proteins are broken down into these

anaerobic organism: an organism that does not need air or free oxygen

anemia: low red blood cell count generally caused by iron deficiency

annelid: one of a phylum of segmented worms, such as the earthworm

artery: blood vessel that carries blood away from the heart

asexual reproduction: reproduction requiring only one parent

atmosphere: thin layer of gases—including nitrogen, oxygen, ozone, and carbon dioxide—that surrounds the earth

auxin: hormone that regulates plant growth

B

bacteria: microorganisms found in water, air, and soil, and within larger organisms

Benedict's solution: chemical indicator used to test for sugar

bile: green liquid produced in the liver, used to emulsify fats

biodegradation: when materials decompose through natural means

biome: region with same type of climate and populations.

biuret solution: chemical indicator used to test for food proteins

blood vessels: a series of tubes in the body through which blood circulates

bromthymol blue: chemical indicator used to test for acids

budding: form of asexual reproduction in which offspring forms from a bud on the parent

C

capillaries: smallest of the blood vessels

capillary action: the process by which, through a combination of cohesion and adhesion, a liquid can rise up through a solid

carbohydrate: energy-rich substance containing carbon, hydrogen, and oxygen; found in starches and sugars

carbon cycle: the cycle of carbon in the earth's ecosystems in which carbon dioxide is fixed by photosynthetic organisms to form organic nutrients and is ultimately restored to the inorganic state by respiration and protoplasmic decay

casein: protein found in milk

cell: basic unit of structure of all organisms

cell membrane: very thin outer skin of a cell

chlorophyll: green material, needed for photosynthesis, found in chloroplasts

chloroplasts: the parts of a plant cell that contain chlorophyll

circulation: the moving of blood through the blood vessels that form the circulatory system

cobalt chloride: chemical indicator used to test for water

(continued)

cohesion: when molecules stick to molecules of their own kind

colloidal solution: fine particles of a substance suspended in another substance

D

deficiency disease: disease caused by lack of a vitamin or mineral in the diet

diet: everything an organism eats

diffusion: the mixing of molecules of one substance with those of another one

digestion: the process by which food is broken down for use by the body

digestive system: system of organs that digest food

diversity: the extent to which there are different species in a particular biome or ecosystem

E

ecosystem: populations in a community, as well as abiotic features with which they interact

emulsification: keeping two liquids that do not combine with each other in suspension, one within the other

endotherm: a warm-blooded animal

environment: the factors and conditions that influence an organism's development

enzyme: chemical that helps food break down during digestion

erosion: the wearing away of soil or rock by wind or water

evaporation: the conversion of a liquid by heat into vapor or steam

exhale: push air out

F

fat: soft, solid organic compound composed of carbon, hydrogen, and oxygen; an essential part of the human diet

fermentation: process in which yeast anaerobically breaks down sugar, producing alcohol and carbon dioxide as by-products

fertilization: when the nuclei of male and female reproductive cells join together

G

gall bladder: sac under the liver where bile is stored

gastric juices: digestive fluids produced in the stomach

gene: a segment of DNA located on a chromosome which directs heredity of specific traits

germination: sprouting of a seed

goiter: swelling of the thyroid gland, caused by iodine deficiency

gravitropism (geotropism): tendency of a plant to grow toward or away from the earth

greenhouse effect: warming of the atmosphere caused by gases reflecting heat back to the earth

H

habitat: a place where an organism lives

hormone: secretion that affects growth and development of an organism

hydrochloric acid: highly corrosive acid found in the digestive juices of the human stomach

(continued)

hydroponics: the science of growing plants in a nutrient solution instead of in soil

hypertonic solution: in cells, solution in which the concentration of dissolved materials is higher outside the cell than the concentration within the cell, which causes the cell to shrink as water leaves it through osmosis

I

indicator: a substance whose physical appearance changes when another specific substance is present; this change shows the presence of the second substance.

indophenol: chemical indicator used to test for vitamin C

inhale: take in air

inherited traits: characteristics passed from parents to their offspring

iris: colored part of the eye that controls the amount of light entering the eye

L

lacteals: parts of the villi through which fats pass into the bloodstream

larva: second stage in the development of an insect, where the young resemble worms

limewater: chemical indicator used to test for carbon dioxide

lipase: enzyme used to break down fats

Lugol's solution: iodine-containing chemical used to test for the presence of starches

M

metamorphosis: marked change in the form of an animal as it develops

minerals: substances found in nature that help the body function and develop normally

mold: simple organism of the fungus kingdom, associated with decay; antibiotics are derived from molds.

molecule: a combination of groups of atoms that are held together electrically

N

nematode: a phylum of small worms that are prevalent in soil

nicotine: stimulant that speeds up heart rate; found in tobacco

O

offspring: new life produced by living things

organism: a living thing

osmosis: type of diffusion where liquid passes through a membrane

oxidation: the process of combining with oxygen

P

pepsin: enzyme found among the gastric juices

peristalsis: muscular movement of the digestive tract

perspiration: sweat; one of the liquid wastes of the body

photosynthesis: the food-making process in plants, using light energy absorbed by chlorophyll, as well as water and carbon dioxide

phototropism: response of a plant to light

preservative: substance added to a food to keep it from spoiling

(continued)

protein: nitrogen-containing nutrient used to build and repair cells

pulse: blood spurt in the arteries

pupa: in metamorphosis, a stage from which the larva emerges as a mature adult

pupil: circular opening at the center of the iris

R

red blood cell: oxygen-carrying cell in the blood

rennin: enzyme found in gastric juices

respiration: inhalation and exhalation of air

response: an organism's reaction to a stimulus

rickets: bone deformation caused by lack of vitamin D

S

saliva: liquid that helps predigest food; secreted by glands in the mouth

scurvy: a disease marked by skin lesions, bleeding gums, and weakness; caused by a deficiency of vitamin C

sexual reproduction: reproduction requiring two parents, male and female

spiracle: tracheal aperture of an insect

spore: reproductive cell

sporulation: producing new spores by the division of older spores

stethoscope: medical instrument used to listen to heartbeats

stimulus: something in the environment that causes living things to respond

stomata: small openings on the underside of plant leaves through which gases pass

T

tracheids: tubular cells in higher plants that serve to transport water

trait: characteristic of living organism

transpiration: the process through which plants emit water vapor into the atmosphere

tropism: movement of a plant toward or away from a stimulus, such as light or gravity

turgor pressure: the pressure that exists inside a cell

V

vegetative propagation: a type of asexual reproduction in which one part of a plant is used to grow a new plant

villi: fingerlike structures in the small intestine through which food is absorbed into the bloodstream

vitamin: nutrient that helps the body regulate growth and development

W

white corpuscle: blood cell that helps fight infection

Y

yeast: single-celled, microscopic fungus